W9-ABS-751

H7
c.1

The Langurs of Abu

Female and Male Strategies of Reproduction

Sarah Blaffer Hrdy

LIBRARY
FLORIDA KEYS COMMUNITY COLLEGE
5901 COLLEGE ROAD
KEY WEST, FL 33040

Harvard University Press
Cambridge, Massachusetts
London, England

For
my mother

Copyright © 1977, 1980 by the President and Fellows of Harvard College
All rights reserved
Second printing, 1980
Printed in the United States of America

Library of Congress Cataloging in Publication Data

Hrdy, Sarah Blaffer, 1946-
    The langurs of Abu.

    Bibliography: p.
    Includes index.
    1. Presbytis entellus—Behavior.   2. Social
behavior in animals.   3. Sexual behavior in animals.
4. Mammals—Behavior.   5. Mammals—India—Mount Abu.
Mount Abu, India.   I. Title
QL737.P93H7   599'.82   80-82476
ISBN 0-674-51057-7 (cloth)
ISBN 0-674-51058-5 (paper)

Preface, 1980

The unusual feature of this book when it first appeared was its focus on female reproductive strategies. The chapters on competition between females and communal "infant-sharing" were, for me, the most exciting. The book devotes equal space to both sexes. It was a disappointment, then—but not a surprise—that most of the interest generated by the book focused on the reproductive strategies of langur males. Male behavior has this power to command (rivet?) attention, partly because adult males are a highly conspicuous minority in most troops of monkeys, but also as a consequence of longstanding traditions within the field of primatology, traditions which are gradually changing.

The propensity of langur males to kill offspring sired by their competitors, as reported in this book, has led to some controversy. Several articles have questioned the hypothesis that destruction of infants by adult males could be adaptive; simultaneously, in a rash of reviews and field reports, much new information about infanticide in other animals has come to light. The value of the criticisms is that they set down in published form widely held assumptions about the "social pathology hypothesis," namely, that behavior disadvantageous to the group or species as a whole cannot evolve and must therefore be maladaptive. Specific objections raised were that (1) infanticidal behavior is poorly documented, and only a handful of eyewitness accounts of the entire sequence of events exist; (2) that infanticide does not necessarily occur every time a new male takes over a troop; (3) that infanticide only occurs in disturbed environments; and (4) that infanticide as a reproductive strategy requires greater calculation than any nonhuman primate could possess (Boggess 1979; Curtin and Dolhinow 1978; 1979; Vogel 1979). The first two points are true. In response, I can only point out that the sparse eyewitness accounts are supported by an impressive amount of circumstantial evidence: case after case of infants attacked by adult males, and of cohorts of infants mysteriously disappearing, just after a new male enters a troop; furthermore, assaults on infants need not be an automatic or universal accompaniment of male take-overs in order to have a genetic base. As to the third point, it is simply not true that infanticide occurs only in disturbed environments. The fourth objection is problematic the moment the evidence from other species is introduced.

Evidence from other species is highly relevant to the infanticide controversy. Field reports described in Chapter 8, as well as very recent studies, indicate that infanticide occurs in a wide variety of primates and other animals living in both disturbed and quite natural environments (Angst and Thommen 1977; Blaffer Hrdy 1979; Butynski 1980; Galat-Luong and Galat 1979; Goodall 1977; Rudran 1979; Struhsaker 1979; Wolf 1980). Recent experiments with rodents indicate that males can in fact gain reproductive advantages from eliminating their competitors' offspring, and that a male with cognitive abilities no greater than that of a lemming or a house mouse is able to discriminate the offspring of an unfamiliar female from offspring possibly his own (Mallory and Brooks 1978; Labov 1980). (These references are listed in an addition to the bibliography on page 328.)

I hope this paperback edition will not only serve as a convenient case study of a particular primate but will also make clear the value of a comparative approach. May it also call attention to a facet of the behavior of female primates too lightly brushed aside the first time around: the extent to which females are competitive creatures of strategy whose preoccupations extend far beyond "mothering" and the traditional boundaries of "maternal behavior."

Acknowledgments

Not acknowledgments, really, but a catalogue of heroes.

As might be expected on the basis of W. D. Hamilton's theory of inclusive fitness, this research received its greatest support from Camilla Davis Blaffer, with whom I share half my genes by direct descent. Daniel Hrdy invested in a current and future spouse by his tolerance and versatile assistance. Odyssean in his photographic, cartographic, editorial, and monkey-watching resourcefulness, Dan helped at Mount Abu—during the time he was able to take off from voyages to the Solomon Islands and life among the shades at Harvard Medical School—and was indispensable.

I first learned of langurs accidentally, while satisfying a distribution requirement in one of Harvard's most popular undergraduate courses, primate behavior, starring Irven DeVore. A remark by the professor concerning the relation between crowding and the killing of infants in this exotic species of monkey brought me back to Harvard a year after graduation to find out why the phenomenon occurred. My first paper in graduate school, entitled "Infant-Biting and Deserting among Langurs," was a less than promising start. The course was meant to be on the evolution of sex differences, but I had attempted to slip in a topic related to my current obsession. According to the grader, an inspired teaching assistant named Robert L. Trivers, the paper had "nothing to do with sex." Later, while I worked under the guidance of (by then) Dr. Trivers, I was to learn how profound the relationship between langur infanticide and reproduction really is.

In the voyage that followed, Professors DeVore and Trivers, together with a synthesizing Olympian, Edward O. Wilson, introduced me to a realm of theory that transformed my view of the social world. These mentors exposed me to the theories of R. A. Fisher, W. D. Hamilton, and G. C. Williams, as well as to their own ideas. Though I make few specific references to them in this book, I hope that my indebtedness will be clear from the tack that my analysis takes.

Sailing between twin dangers of group selection and the naturalistic fallacy, I was fortunate in the company aboard. Many of the ideas presented here were initially tried out upon, or pirated from, my friends: Peter Ellison, Martin Etter, John Fleagle, James Malcolm, Nancy Nicolson, Joe Popp, Allen Rutberg, Jon Seger, Pat Whitten, and Richard Wrangham. In the realm of langurs, I am deeply indebted to the Indian primatologist S. M. Mohnot, who has observed langurs longer than any person in the world. During my first weeks in India, Dr. Mohnot taught me the habits of these monkeys, and it was at his suggestion that I went to Mount Abu. The hospitality of S. M., of his wife, Uchhab, and of Professor M. L. Roonwal provided a stimulating and pleasant oasis during two visits to the desert-dwelling langurs of Jodhpur. In addition, I thank Naomi Bishop, Suzanne Ripley, and Yukimaru Sugiyama for so generously sharing their knowledge of langurs with me.

For comments on the manuscript I thank Naomi Bishop, John Fleagle, Glenn Hausfater, Robert Hinde, Alison Jolly, Mel Konner, Suzanne Ripley, Peter Rodman, and Jon Seger. Jon Seger also came to my rescue as I wrestled with the problem of the extent to which individuals in langur troops are related to one another. For him, the bowstave bent easily, and his results are presented in appendix 3.

I thank Virginia Savage for the Circean skill she used to transform photos into drawings, and John Melville Bishop for translating color photos into black and white. K. M. Dakshini and Steve Berwick identified plant specimens. B. Beck, D. Ben Shaul, N. Bishop, J. Curtin, J. Fleagle, D. Fossey, S. Kitchener, J. McKenna, S. M. Mohnot, J. Oates, J. Oppenheimer, S. Ripley, T. Struhsaker, R. Tilson, C. Wilson, and K. Wolf provided valuable unpublished information. To the citizens of Mount Abu who tolerated my trespassing on their backyards in pursuit of langurs, and most especially to Aftab and Mona Ali and Nirmal Kumar Dhadhal and their families, I extend my thanks for their hospitality to a stranger in a foreign land. On home ground, I thank the Peabody Museum and its staff for providing me with a base and for assistance in many ways.

The penultimate acknowledgment is reserved for a different species of hero: *Presbytis entellus*, the hanuman langur of North India, named for Hanuman, the Hindu monkey-god, and for Entellus, the Sicilian boxing champion lionized in the *Aeneid*. In fact, heroes abound in the taxonomy of these aristocratic silver monkeys: *Presbytis entellus achilles*, the race of the high Himalayas, *P. e. ajax* of the Northern Punjab, *P. e. anchises* of the vast Deccan plateau, down to *P. e. thersites* of Sri Lanka, somehow named for a Greek fighter notorious for his ugliness, deformity, and foul mouth. Anyone heroic enough to read on to the end of this book will learn why the identification of langurs with warriors was an appropriate taxonomic choice, and why the final salute must be to the prescience of the nineteenth-century British naturalists who first went out to study the Hanuman.

S.B.H.

Contents

Figures

Color Figures
(between pages 140 and 141)

Note: Unless otherwise indicated, all photographs are by D. B. Hrdy and
the author.

Tables

The Langurs of Abu

"You were talking about sex-appeal a moment ago . . . "

"I never used any such expression. I said sexual selection. It gives the best results in evolution."

EDWIN CERIO,
Love Story of the Blue Lizard (1944)

1 / Opponent Consorts

Titus Andronicus has been one of Shakespeare's least popular plays and is rarely performed today. Few people are familar with its intricate and grisly subplots, such as the tale of Aaron the Moor. Shortly after Aaron's mistress marries the emperor of Rome, she gives birth to a "blackamoor" child. Alas, the exotically hued infant stands as a tattletale genetic marker of the empress's infidelity. Given the circumstances, the nursemaid's lament is understandable: "Here is this babe," she moans, "as loathsome as a toad amongst the fair faced breeders of our clime." The empress's future is in jeopardy; she begs Aaron to destroy their son before her husband hears the news. Heedless of the empress's dilemma, Aaron refuses, threatening to skewer anyone who touches his firstborn son, for as Aaron so clearly perceives, "My mistress is my mistress, this [babe] my self . . . this, before all the world will I keep safe, or some of you shall smoke for it in Rome."

Small wonder that a play replete with infanticide, dismemberments, rape, and cannibalism has not weathered well the test of time. To modern audiences, these events seem implausible and unnatural. Nevertheless, if more primatologists had seen this play before going off into the field, they might better have understood the behavior unfolding before them in the savannas and forests where monkeys are studied.

The conflict of interest between Aaron and his mistress is basic to all sexually reproducing creatures where the genotypes—and hence self-interests—of two consorts are necessarily not identical. The survival of the empress and her previous children (who are prisoners of war kept by the emperor) depends upon her continued marriage to her powerful husband. The illegitimate infant, however, may be Aaron's sole opportunity to sire a son. It is to the advantage, then, of one consort, but not the other, for this child to live.

Nowhere among the primates has such a conflict of interest between consorts been more clearly documented than among hanuman langurs, a species in which adult males routinely resolve their conflicting interests with females by killing the females' offspring. But infanticide is only one instance, albeit a particularly striking one, of the selfish behaviors found among langurs. Hanuman langurs in a variety of social, and antisocial, ways are unequivocally self-serving.

Competitive and cooperative behaviors of male and female hanuman langurs can be analyzed in terms of how the behaviors contribute to, or detract from, the fitness—measured solely in terms of surviving offspring—of the individuals concerned. Individual fitness, however, is only part of the story. Animals may achieve genetic representation in subsequent generations through an alternate route as well, by behaving in such a way as to increase the fitness of close relatives who share genes by common descent. This indirect route to genetic representation, first described by W. D. Hamilton in 1964, is known as "kin selection," and is based on the concept of "inclusive fitness." Inclusive fitness refers to the sum of the individual's own fitness, plus the effects that his behavior has on the fitness of his relatives, and vice versa. The interplay between individual and inclusive fitness, selfishness and cooperation, lies at the heart of understanding langur social life. A knowledge of the langur breeding system, especially the fact that females will usually only be distantly related to their male consorts, leads to a preliminary understanding of the tensions that characterize relations between the sexes in this species.

The universality of conflict between the sexes is one reason that langurs might be of interest to human readers. A second reason is that the history of langur studies provides an obvious but all too relevant paradigm of our efforts to understand ourselves and creatures like us. To tell this tale, in the next chapter I shall review what has been known—and also what has been dismissed as knowledge—about langurs for a long time.

2 / Langurs through the Eyes of Humans

Many of the [Adelie penguin] colonies are plagued by
little knots of "hooligans" [peripheral males] who hang
about their outskirts, and should a chick go astray it
stands a good chance of losing its life at their hands.
The crimes that they commit are such as to find no
place in this book.

G. Murray Levick, *Antarctic Penguins* (1915)

Ranging from altitudes near 4,600 meters in the Himalayas to near
sea level and living in habitats that range from moist montane forest
to semidesert, populations of the black-faced, gray-furred common
or hanuman langur occur in pockets and in swaths from Nepal down
through India, including both sides of that vast subcontinent, southward to the island country of Sri Lanka. The name "langur" derives
from the Sanskrit *langulin*, "having a long tail," the epithet
"hanuman" from the monkey-god and loyal servant to King Rama,
the legendary Hanuman who retrieved Rama's stolen wife from the
lecherous king of Lanka (fig. 2.1). Both names bespeak a history of
thousands of years during which hanuman langurs have lived in
close association with man. Because of their status in Hindu myth-
ology, monkeys are tolerated by Hindus throughout India, and
langurs can still be found ranging freely in towns and in some major
cities. In 1972, for example, while passing through the congested
industrial city of Ahmedabad, I was astonished to see a band of
langur males zigzag through traffic, ricochet off moving three-
wheeler scooter-taxis, and then disappear out of sight.

The first published reports of the behavior of hanuman langurs
are accounts of early British naturalists. In 1836 a description of a
band of males who invaded a bisexual troop of langurs appeared in
the August issue of the *Bengal Sporting Magazine* as a "story of the

3

Fig. 2.1. Hanuman the monkey-god and loyal servant to King Rama, whose exploits were compiled in the epic *Ramayana* sometime around the fifth century B.C. Hanuman was sent to the island kingdom of Lanka to rescue Rama's stolen wife, Sita. Though Hanuman was able to locate the captive queen, she doubted his intentions. To prove his loyalty, Hanuman ripped open his chest to display what was engraved there upon his heart: King Rama, Sita herself, and Rama's brother. Shortly thereafter the queen's rescuer was discovered by the king of Lanka. In retaliation, the king tied burning faggots to Hanuman's tail. Undaunted, Hanuman raced around the island setting the whole kingdom afire. Then, Hanuman doused his flaming body in the Indian Ocean and leapt back across the channel to India. (Based on a folk representation of Hanuman painted onto a pipal leaf.)

stronger sex trying conclusions amongst themselves for the charms of the gentler one." According to this rather prescient, pre-Darwinian account:

> The males are exclusively the combatants and the strongest usurps the sole office of perpetuating his species through the reciprocal agency of his female associates . . . At a particular season of the year the great body of he-monkies which had been leading a monastic life deep in the woods, sally forth to the plains and mixing with the females a desparate conflict ensues for the favours of the fair lady pugs . . . This continues for several days at the end of which time, one male more valorous than the rest will be found in possession of the flock . . . [after which] . . . a kind of conference takes place, the females delivering up their half-grown male offspring to the care of the [defeated males] who troop away to the jungle re-inforced by hopeful juniors [quoted in Hughes, 1884].

Similar accounts were published in the *Proceedings of the Asiatic Society* (Hughes, 1884) and in the *Journal of the Bombay Natural History Society* (J. F. G., 1902). Hughes witnessed combatant males murder one another and (if we are to believe the account) females in the troop attack and castrate an invading male.

> In April 1882, when encamped at the village of Singpur in the Sohagpur district of Rewa state, my attention was attracted to a restless gathering of "Hanumans" . . . two opposing troops [were found] engaged in demonstrations of an unfriendly character.
> Two males of one troop . . . and one of another—a splendid looking fellow of stalwart proportions—were walking round and displaying their teeth. The solitary gladiator headed a much smaller following than that captained by the other two, and strange to say, instead of the whole number of monkeys joining in a general melee the fortune of the question that had to be decided appeared to have been intrusted to the representative champions.
> It was some time—at least a quarter of an hour—before actual hostilities took place, when, having got within striking distance, the two monkeys made a rush at their adversary. I saw their arms and teeth going viciously, and then the throat of one of the aggressors was ripped right open and he lay dying. He had done some damage however before going under, having wounded his opponent in the shoulder . . .
> I confess that my sympathies were with the one champion who had gallantly withstood the charge of his enemies; and I fancy the tide of victory would have been in his favour had the odds against him not been re-inforced by the advance of two females . . . Each female flung

herself upon him, and though he fought his enemies gallantly, one of the females succeeded in seizing him in the most sacred portion of his person, and depriving him of his most essential appendages. This stayed all power of defense, and the poor fellow hurried to the shelter of a tree where leaning against the trunk, he moaned occasionally, hung his head, and gave every sign that his course was nearly run . . . before the morning he was dead [1884:148-149].

Although extraordinary events were remembered and set down by travelers, by and large the langur literature of the day was dominated by taxonomic studies which were at that time, and remain today, a "bewilderment to all but the specialist" (Pocock, 1934). The problem was not only the independent personalities and lack of communication between naturalists but also the monkeys themselves, a widespread assemblage characterized by much intergradation between subspecies, a large degree of overlap of external traits, as well as individual variability in skulls (Pocock, 1928a, b; 1934; and Hill, 1939). Interspersed between taxonomic debates, a few encounters between langurs and their predators were also described, especially the curious practice langurs have of descending to the ground when threatened by leopards or dogs (for example, Champion, 1928, cited in Nolte, 1955; Brander, 1939; and Pythian-Adams, 1940).

The first person actually to study the social organization of Asian monkeys in their natural habitat was Charles McCann, a botanist (1938). Inspired by the publication of Sir Solly Zuckerman's The Social Life of Monkeys and Apes (1932), McCann undertook a more detailed examination of langur life in which, at a very early date for primate studies (1933a), he set down many of the significant features of langur sociobiology: the capacity of langurs to go for long periods without water; the presence of a breeding season and a birth peak; the fluctuation in the number of adult males present in bisexual troops; male take-overs; territoriality; the rescue of an infant by an adult male langur from the same troop; the harassment of copulating couples by other adult females in the troop; postconception estrous solicitations of males by pregnant females; the return of older infants to suckle after the birth of a new sibling; the changes with age in infantile coat color; and the peculiarly colobine trait of postnatal infant transfer. In a perceptive statement in 1933, McCann summed up the main organizing principle of langur social life: "The Hanuman is a true polygamist . . . A single male tries to take charge of . . . all the females of the troup and establishes himself as their overlord by driving out all the potent males, big or small." In addition, McCann published brief accounts of closely related species—the capped langur (*Presbytis pileatus*), the Nilgiri langur (*P. johnii*)—as well as the hoolock gibbon (*Hylobates hoolock*). Though rarely read today, McCann's pioneering study of langurs foreshadows many interpretations of their behavior that will be offered here.

Despite the shrewdness exhibited by some of the early naturalists, one of the first steps of modern primatology was to put aside these "anecdotal" accounts. Unfortunately, in the case of langurs, this policy has meant a long detour in our understanding of why langurs behave as they do.

By the late 1950s the modern era of primate studies—launched primarily by social scientists—had begun. The worldview of these workers was profoundly influenced by current social theory, in particular by the Radcliffe-Brownian view that social organization was a "functionally integrated structure" (Jay, 1963c:239). Not surprisingly, the early reports of primate societies uniformly described monkeys maintaining complex social organizations in which each had a role to play in the life of the group and all members functioned together to ensure the group's survival. If among vervet monkeys, for example, young males were routinely found at the outskirts of the troop, it was said that these immature animals, the most expendable of the troop's resources, were there to act as a buffer and to be caught in case a predator should appear—hence protecting the more valuable members of the troop. If among baboons dominant males seemed to have greater access to estrous females, this was said to ensure that only the best animals mated, thereby improving the genetic stock of the group as a whole. The question was rarely asked how such behavior might benefit the individual concerned, or how a more powerful individual might benefit from forcing other animals to comply to his advantage. Typically, behavior was explained in terms of group survival.

Langurs are among the most terrestrial and observable of primates. Accordingly, *Presbytis entellus* was one of the first species to be intensively studied. In 1958-59, Phyllis Jay (now Phyllis Dolhinow) undertook her pioneer research of langurs at the game preserve of Orcha, in Central India, and near the village of Kaukori on the cultivated Gangetic plain of North India. In her summation of langur social organization she wrote: "A langur troop is not merely a formless agglomeration of individuals, but is a unit with a definite shape or structure based on intricate patterns of social relationship among the members. Every animal in the troop contributes to the maintenance of this structure by participating in these relationship patterns in a certain characteristic manner" (Jay, 1963c:223). This report illustrates the widespread conviction of the times that animals behave as they do in order to maintain, not disrupt, the prevailing social structure. In line with this outlook, the early naturalists' reports of fighting among langurs were dismissed as "anecdotal, often bizarre, certainly not typical behavior" (Jay, 1963c:8). Based on her own observations at Kaukori, Jay was convinced that McCann's 1933 report had "greatly exaggerated both the aggressive and the sexual behavior which he observed" when describing the "overlord male" as driving out all the other males from the troop during the period of sexual receptivity. In contrast, Jay emphasized the very peaceful nature of langur life: "Relations among adult male langurs are relaxed. Dominance is relatively unimportant . . . Aggressive threats and fighting are uncommon . . . the relaxed nature of langur life is one of the first characteristics an observer notices" (1965).[1]

But a second langur study, carried out in 1961, forced reconsideration of the question of aggressiveness among langurs. For two years, five members of the Japan India Joint Project tracked the movements and activities of troops of langurs in the fields and teak forests surrounding the village of Dharwar in South India. In June of 1962, Yukimaru Sugiyama watched as a band of seven langur males drove out the leader of a bisexual troop, after which one of the band members usurped sole control by driving away the other males. All six infants in the troop were then bitten to death by the new leader (1965). In the course of the project, six additional take-overs were witnessed by Sugiyama and his co-workers, and four other take-overs were inferred by changes in troop compositions (Yoshiba, 1968). Curiously, and contrary to previous reports of the solicitude of langur mothers, some of the mothers at Dharwar whose infants were attacked by invading males abandoned their offspring prior to death.

How could these reports be compatible with the notion, which has since become entrenched in the primate literature, that langur social relations are among the most "relaxed" and "nonaggressive" to be found among primates? How could infanticidal behavior be compatible with the good of the group?

In additon to the belief that social creatures behave as they do to ensure the group's survival, two other viewpoints popular in the

---

1. This interpretation was not entirely consistent with Jay's additional finding that some langurs bore scars as "silent testimony of past fights." In four months of observation at Kaukori, adult males were wounded on five occasions (1963c:184).

1960s may have contributed to the notion that under normal conditions murder could not occur among langurs. The first of these was the view that only man routinely murdered members of his own species. Our uniquely murderous reputation was based on the optimistic conviction that none of nature's creatures unconfused by culture could behave in such a way as to threaten the survival of their species. Perhaps the most distinguished proponent of this view is Konrad Lorenz. Though Lorenz was aware that on occasion a lion or wolf might kill a member of its own kind, he regarded such cases as rare exceptions. In his classic (1966) study of animal and human aggression, Lorenz insisted that any species capable of efficiently murdering a conspecific must possess "sufficiently reliable inhibitions which prevent self-destruction of the species" (1971 edition). Less lethally equipped creatures such as doves, hares, or chimpanzees, he argued, needed no such inhibitions because the possibility of such an animal seriously injuring one of its own kind would so rarely occur in nature.

With the dawning awareness that very high population densities can have adverse consequences on behavior, a second reason emerged for regarding the report from Dharwar as a description of abnormal behavior. Langur population densities at Dharwar were as much as 44 times higher than those of the peaceful langurs at Jay's study site, and this fact did not escape the attention of scientists attempting to explain the extremes of aggression witnessed in South India. A recent experimental study of rats by John Calhoun (1962) had shown that at very high densities normal social conventions could break down. Under extremely crowded conditons, rats sank into a pathological state, a "behavioral sink" characterized by excessively high infant mortality resulting from inadequate maternal care, infanticide, and cannibalism. The fact that some individuals (namely some harem-holding males) continued to prosper even under conditions Calhoun described as a "behavioral sink" was not stressed at the time. His emphasis was on social pathology, not on selfish strategies that remained adaptive for those individuals able to adopt them.

Extrapolating from such studies, a number of researchers suggested that severe crowding could transform usually relaxed relations among langurs into abnormally aggressive ones (Dolhinow, 1972; see also Sugiyama, 1967, cited in Crook, 1970; Bygott, 1972; Eisenberg et al., 1972; Rudran, 1973b). Infanticide among langurs was dismissed as an instance of "social pathology" or some other "dysgenic" behavior (Warren, 1967).

It was to test the hypothesis that crowding was responsible for infant killing among langurs that in 1970 I decided to go to India. By the time I finally arrived, in June of 1971, a new report of infanticide among langurs, this time from Jodhpur in the desert regions of Rajasthan, far to the north of Dharwar, had just been published (Mohnot, 1971b). Already it seemed possible that infanticide might turn out to be a far more widespread and normal behavior than my hypothesis in hand suggested.

By the time I concluded my research, four years later, I had re-learned what a few British naturalists had known almost a century before: that langur males compete fiercely for possession of females, and that in the process, conspecifics are sometimes killed. Furthermore, langurs are far from unique in this respect. A host of species has been recently added to the list of creatures known to kill conspecifics for motives other than eating them. These include such diverse groups as lions, hippos, bears, wolves, wild dogs, hyenas, rats, rabbits, lemmings, herring gulls, storks, European blackbirds, eagles, and more than fifteen types of primates—or sixteen, counting man. Some of these appear to be directly comparable to the hanuman langurs.

Far from being maladaptive, infanticide was found to be a wide-spread adaptation to normal conditions of langur life that was quite advantageous to those males who practiced it. In chapter 8, I will detail the various lines of evidence which led to this conclusion. The question that remained was not why langurs kill infants, but why we had for so long chosen to regard these incidents as unnatural or pathological behavior.

The answer, of course, is that nowhere in the "group maintenance" or "species survival" scheme is there room for the antisocial behavior that arises as a by-product of sexual selection, and which in the langur case has taken an extreme form. Nor is there room for some of the other unabashedly selfish behaviors that I only began to understand in the last years of my study. These included a hierarchical system of female dominance in which older animals are pushed aside by their younger female relatives, and a pattern of infant caretaking that often results in cavalier abuse of their charges by females other than the mother.

In the epigraph introducing this chapter, an early Antarctic explorer admits publicly to euphemizing. Confronted with a discrepancy between what was and what ought to be, he chose to delete the misdemeanors of hooligan penguins (in that instance, probably cannibalism). The langur whitewash has been more subtle but in the same anthropocentric vein. Because we are human and hold ourselves to be unique, we tend to value highly our own ideals.

Not surprisingly, when we first began to intensively study our closest nonhuman relatives, the monkeys and apes, an idealization of our own society was extended to theirs: thus, according to the first primatological reports, monkeys, like humans, maintain complex social systems geared towards ensuring the group's survival. It is this peculiar misconception about ourselves, and about primates, that lends the history of langur studies its significance. By revealing our misconceptions about other primates, the langur saga may unmask misconceptions about ourselves.

MONKEYS CONSTRUCTING THE BRIDGE AT LANKA.

3 / A Highly Adaptable Colobine

A creature about the size of a springer spaniel, endowed with the slender-waisted elegance of a greyhound, the hanuman langur is the most terrestrial of all the Colobinae, a far-flung subfamily of African and Asian monkeys. The hanuman langur is also the most widely spread of any other single species within this vast subfamily, and can be found in extraordinarily varied ecological circumstances. But as with any creature, there are limits to diversity. Langurs' unique package of physiological and behavioral capacities not only opens options but imposes constraints on the patterns of survival that are possible. Similarly, the basis for all langur social behavior is phylogenetic legacy tempered by current environmental pressures. Information on the ecology and behavior of langurs from a diverse array of study sites in India, Nepal, and Sri Lanka allows us to construct a model of langur social organization, and to distinguish between langur behaviors that are relatively constant and those that vary with habitat and demographic conditions. First, however, it would be useful to examine the history of colobines as a whole.

Given a Colobine

Old World monkeys are divided into two subfamilies, the Cercopithecinae, comprising baboons, macaques, and guenons, and the Colobinae, composed of colobus monkeys, langurs, and a group known as the odd-nosed monkeys of Asia. A closer look at the whole array of colobines and a bit of speculation about the common ancestor of these forms will allow us to establish an evolutionary context for understanding the behavior of langurs. Few tasks are plagued by greater uncertainties than the reconstruction of the life-styles of an extinct ancestor.

One way to reconstruct an ancestral form of a particular group is to examine geographically disparate members of the group and to ascertain which features they share in common: such features are likely to have been present in their most recent common ancestor. All colobines, for example, resemble each other in two respects. Unlike other Old World monkeys, colobines lack cheek pouches for the temporary storage of food. On the other hand, they have large complex stomachs that allow them to digest quantities of mature leaves. Colobines are often called "leaf-eaters" or "leaf monkeys" for this reason. Among *Presbytis entellus,* the special diverticular form of the stomach permits ruminant-like bacterial fermentation involving large numbers of anaerobic bacteria (Bauchop and Martucci, 1968). Given the omnipresence of the leaf-eating adaptation among colobines, it seems safe to assume that a similar adaptation existed in their common ancestor. Similarly, most colobines differ from other Old World monkeys in that the females lack sexual swellings. Whereas among baboons or chimpanzees sexual receptivity is signaled by vivid red swellings in the perineal region, sexual receptivity among langurs and other colobines is indicated largely by behavior. Again, it seems likely that the common ancestor also lacked sexual swellings or was at least predisposed to their loss.

A second way to reconstruct the ancestral colobine is to examine the fossil record directly. Colobines probably originated in Africa over twelve million years ago and migrated across Europe into Asia (Simons, 1972). The earliest fossil evidence for members of this subfamily comes from early Pliocene deposits in Eastern Europe and the Middle East. The cranium of this fossil form, known as *Mesopithecus,* resembles that of the African colobines, but its faunal associations suggest that *Mesopithecus* lived near open country and grasslands, pursuing a more terrestrial way of life than any of its African relatives do today. This impression is strengthened by *Mesopithecus'* limb proportions, which resemble those of the hanuman langur, by far the most ground-adapted of all extant colobines (Delson, 1973). Numerous skulls and postcranial bones have been found in Pliocene deposits in Greece; they indicate a sexual dimorphism among the fossil forms much greater than among current African colobines but similar to some of the Asian colobines in which males are substantially different in size and morphology from females (fig. 3.1). These resemblances between the fossil forms and contemporary Asian colobines lend support to Simons' view that *Mesopithecus* is near the "basal ancestry" of Asian colobines.

Late Pliocene deposits, about four million years old, yield the next clear evidence for the subfamily and also for the split between African and Asian forms which had taken place by that time. *Paracolobus*, discovered in Kenya by Richard Leakey, shows clear affinities with contemporary African colobines. The Asian colobine of about the same period, represented in fossil finds from the Siwalik Hills of India, is known as *Semnopithecus*, a term later supplanted by the modern designation *Presbytis*. This fossil form, described by Lydekker (1885), closely resembles Indian langurs of today.

The task of reconstructing an extinct ancestor becomes even more difficult when we shift from morphology to behavior, since

Fig. 3.1. Among the sexually very dimorphic proboscis monkeys of Borneo, females are smaller than males and have small, uptilted noses. Males weigh more, and have enormous drooping snouts that continue to grow throughout adulthood.

there is only limited evidence from living species to guide us (table 3.1). Leaving aside the controversies that surround the taxonomic classification of colobines (recently reviewed by Groves, 1970; Medway, 1970), and subscribing to names in prevalent usage (table 3.1), the extraordinarily diverse living members of the colobines can be usefully divided into three clusters: African leaf-eaters of the genus *Colobus* (including two subgenera: *Colobus* and *Procolobus*); Asian leaf-eaters of the genus *Presbytis*; and the odd-nosed monkeys, which include members of four genera found only in southern China and southeast Asia: *Rhinopithecus*, *Pygathrix*, *Nasalis*, and *Simias*. Of the Asian colobines, only four species—the hanuman langur, the Malaysian lutong, the Nilgiri langur of western India, and the purple-faced leaf monkey of Sri Lanka—have been the subjects of long-term behavioral studies. In fact virtually no in-habitat information exists for six species within the subfamily, including two whole genera, *Pygathrix* and *Rhinopithecus*. Another Asian colobine, the golden langur of Assam, was only discovered as recently as 1955 and along with most other colobines has scarcely been studied.

A survey of the better known colobines reveals a complex of behavioral features shared by members of the subfamily in Africa, Asia, and Southeast Asia. Close similarities between species that in other respects (geographical distribution and appearance) diverge so widely suggest that similar traits may have occurred in the most recent common ancestor of these forms. Some of these behavior patterns may be part of the leaf-eater adaptation of colobines. Some leaf-eaters live year-round in a fixed area despite seasonal aridity. In the cases of the hanuman langur and the black-and-white colobus monkey of East Africa, territorial behavior may be a corollary of living year-round in a circumscribed home range with occasionally scarce resources (Clutton-Brock, 1974). Whether or not this correlation holds for other members of the subfamily, territoriality and especially the paraphernalia of intertroop spacing such as long-distance male vocalizations, and highly visible male displays are widespread among the Colobinae. Typically, these species combine dramatic leaps through the treetops and branch shaking with roars (among East and West African colobus monkeys), brays (Malaysian leaf-eaters), reverberating honks (proboscis monkeys), or the eerie "whoop wha-oop" of the hanuman langur.

Among the most striking of the resemblances between hanuman langurs and other colobines is the recurrence within the subfamily of infanticide. The occurrence of male take-overs, widely documented for hanuman langurs, has also been reported among purple-

TABLE 3.1. The subfamily Colobinae.

Species	Common name	General location	Behavior studies
Colobus polykomos	King colobus	Africa	Sabater Pi, 1973
Colobus guereza (also abyssinicus)	Black-and-white or Abyssinian colobus	Africa	Schenkel and Schenkel-Hulliger, 1966; Wooldridge, 1969; 1971; Marler, 1969; Leskes and Acheson, 1971
Procolobus badius	Red colobus	Africa	Clutton-Brock, 1974; 1975; Struhsaker, 1975
Procolobus verus[a]	Olive colobus	Africa	Booth, 1957
Procolobus kirkii[a]	Kirk's colobus	Zanzibar	
Presbytis senex[b]	Purple faced leaf monkey	Sri Lanka	Rudran, 1973a, b; Manley, in preparation
Presybtis johnii	Nilgiri langur	Western India	Poirier, 1970
Presybtis entellus	Hanuman langur	Indian subcontinent	(see table 3.3)
Presbytis geei	Golden langur	Assam	Gee, 1955; 1961; Mukherjee and Saha, 1974a
Presbytis pileatus[a]	Capped langur	Assam	McCann, 1933b; Tilson, personal communication
Presbytis obscurus	Dusky leaf monkey	Malaysia	Badham, 1967; Horwich, 1974; Hunt-Curtin, in preparation
Presbytis cristata	Silvered leaf monkey or lutong	Malaysia, Sumatra, Borneo	Bernstein, 1968; Wolf and Fleagle, in press
Presbytis melalophos	Banded leaf monkey	Malaysia, Sumatra, Borneo	Harrisson, 1962; Hunt-Curtin, in preparation
Presbytis potenziani	Mentawei leaf monkey	Mentawei Islands	Tilson and Tenaza, 1976

Species	Common name	General location	Behavior studies
Presbytis aygula	Sunda Island langur	Java, Sumatra, Borneo	Wilson and Wilson, in press
Presbytis frontatus[a]	White-fronted leaf monkey	Borneo, Sarawak	Rodman, 1973
Presbytis rubicundus[a]	Maroon leaf monkey	Borneo, Sarawak	Stott and Selsor, 1961
Presbytis phayrei[a]	Phayre's leaf monkey	Burma, Laos, Cambodia	
Presbytis francoisi[a]	François' leaf monkey	Southern China	
Rhinopithecus roxellanae[a]	Golden monkey	Western China	Allen, 1938
Rhinopithecus avunculus[a]	Tonkin snub-nosed monkey	Central China, North Vietnam	
Pygathrix nemaeus	Douc langur	Vietnam	Hill, 1972; Lippold and Brockman, 1974; Kavanagh, in press; Lippold, in press
Nasalis larvatus	Proboscis monkey	Borneo, Sarawak	Kawabe and Mano, 1972; Kern, 1964a
Simias concolor[a]	Pagai Island langur	Mentawei Islands	Tilson, in preparation

a. Virtually no behavioral information available.
b. According to Groves (1970), P. senex and P. johnii are probably conspecifics.

faced leaf monkeys in Sri Lanka (Rudran, 1973b), the Malaysian lutong (Wolf and Fleagle, 1977), and the black-and-white colobus of Uganda (J. Oates, personal communication). In two of these four species (Presbytis entellus and P. senex), adult males entering from outside routinely kill unweaned infants. Both Fleagle and Oates reported the disappearance of infants after take-overs, suggesting that infanticide may also occur among lutongs and black-and-white colobus monkeys. The suspicion of infanticide among wild colobus is further strengthened by a report of an adult male black-and-white colobus killing infants sired by another monkey after he replaced the previous male in a caged group at the Lincoln Park Zoo in Chicago (S. Kitchener, personal communication). On the basis of troop compositions recorded by Frank Poirier, Rudran (1973b) has also inferred the occurrence of infanticide among Nilgiri langurs (according to Groves, 1970, a conspecific of *P. senex*). Among the little known Mentawei leaf monkeys, a four-day old infant disappeared rather suspiciously just hours after its father was replaced by a new male (R. Tilson, personal communication).

Adult male infanticide has been reported for a variety of primates, and the list appears to be growing as more species are observed for longer periods. Nevertheless, in proportion to the number of man-hours that colobines have been watched, infant-killing is surprisingly prevalent among members of this subfamily compared to other Old World monkeys.

After a take-over, a single male ousts all other adult and subadult males. Hence, in areas where take-overs are common, the one-male pattern of troop structure predominates. Broadly speaking, however, colobines are flexible in regard to the number of adult males that may be present in a troop. Tolerance between adult males is highly dependent on the situation. Both one-male and multimale troops have been reported for almost all of those colobine species for which information is available from several sites.[1] There may, however, be a tendency towards one-male harems.

Perhaps the single feature that most distinguishes this subfamily from other primates is the flamboyant coloration of newborn infants. Natal coats dramatically different from those of conspecific adults are a colobine hallmark. Black-and-white colobus mothers give birth to snow-white babies (fig. 3.2). Several species of dusky grey Malaysian leaf-eaters produce brilliant orange offspring with startling circles of white about their eyes. Elsewhere in the subfamily, newborn infants have pale coats with two dark streaks crisscrossing their back in the "cruciger" pattern typical of *Presbytis melalophos*, *P. aygula*, and their near relatives. Similarly, the rare douc langurs of Vietnam are born a chipmunk-colored chestnut with a dark stripe down the spine.

In addition to their flamboyance, colobine newborns are extra-

---

1. Species in which both one- and multimale troops have been observed include *P. melalophos* (Harrison, 1962; Hunt-Curtin, in preparation); *P. cristata* (Bernstein, 1968; J. Fleagle, personal communication); *Colobus guereza* in Uganda and Ethiopia (Marler, 1969; Dunbar and Dunbar, 1974b) and *P. entellus*.

Fig. 3.2. In the vulnerable months just after birth, this newborn black-and-white colobus will wear a snow-white coat markedly different in color and design from the white-mantled, white-bearded pelage of the adults.

ordinarily attractive to other troop members, and mothers freely give up their infants to conspecific caretakers soon after birth. Postnatal transfer of infants occurs earlier among the Colobinae than any other group of primates—except of course ourselves.[2] A

2. Outside of the Colobinae, early transfers have also been reported for wild vervet monkeys (at fourteen hours by Struhsaker, 1967); for gelada baboons (at day four by Dunbar and Dunbar, 1974a); for Barbary macaques (by day three to an adult male; Burton, 1971); and for Lowe's guenon (at day ten by Bourlière et al., 1970). Generally speaking, however, the Cercopithecinae are characterized by more possessive mothers who refuse to give up their infants in the first weeks or months of life (for example, baboons, patas, various macaque species). In some matrifocal species such as the chimpanzee, exceptions occur when siblings are allowed to inspect and even to hold infants prior to the time that infants are generally available to other troop members.

hanuman langur infant at the San Diego Zoo was first transferred just sixteen minutes after birth (McKenna, 1974a), and among caged black-and-white colobus, first transfer was observed within seven hours (Wooldridge, 1969). Similarly, in the wild, langur mothers give up their infants to other females within hours of parturition, and infants may spend up to 50 percent of their first day of life away from their mother. Early transfer of newborns to other troop members, usually females and frequently nulliparous females, has been reported for lutongs, Nilgiri langurs, caged spectacled leaf monkeys, and caged douc langurs (Hill, 1972). Among black-and-white colobus, males may also play an active role in caretaking (Wooldridge, 1971).

Almost certainly relaxation of predation pressures has given arboreal colobines a certain leeway in evolving conspicuous natal coats. Yet even these tree-dwellers are occasionally caught by raptorial birds or felids.[3] Because there are potential disadvantages to a flamboyantly colored infant which is conspicuous to predators as well as to conspecifics, one would expect that there must be compensating advantages for an infant which attracts attention.

These advantages may include rescue of an endangered infant by caretakers other than the mother, as well as care for the infant while the mother forages (Blaffer Hrdy, 1976). In the more terrestrial, and presumably more vulnerable, colobine species, infant garb is more discreet but decidedly distinct from adult pelage. Infants among hanuman langurs are all black and closely resemble the dark infants characteristic of savanna-dwelling Cercopithecinae, such as baboons and vervets. The partially terrestrial proboscis monkey of Borneo has evolved a different sort of compromise to limit the message broadcast by natal flamboyance. Only the newborn's bright blue face is strikingly different from adult coloration.

Not all colobines wear distinctive coats at birth. Among the olive colobus monkeys and perhaps the snub-nosed langur of China, the coat color of infants resembles that of the adult (W. W. Howells, personal communication). Interestingly, the olive colobus is the only primate species in which the mother is reported to carry her newborn infant in her mouth, as a cat would a kitten (Booth, 1957). Also unusual among colobines are the red colobus, who do not pass new-

3. See Booth's account (cited in Jolly, 1972a) of a colobus infant snatched away by a monkey eagle. Raptorial birds have also been known to threaten *P. entellus* and *P. cristata* infants (Rahaman, 1974; J. Fleagle, personal communication).

borns among troopmates. Along with the olive colobus, the red colobus exhibits sexual swellings (absent in every other colobine studied) and both have somewhat different jaw musculature from other colobines (Kuhn, cited in Groves, 1970). The red colobus is odd man out in yet another respect; instead of the typically colobine harem, members of this species travel in large, multimale groups. This may also be true of *Rhinopithecus*. When last described (Allen, 1938), these burly snub-nosed monkeys were said to roam parts of China in groups of one hundred or more.

Among the strangest colobines studied to date, however, are those of the Mentawei Islands, off the west coast of Sumatra. Recent field studies of two colobine species living there, *Presbytis potenziani* (Tilson and Tenaza, 1976) and *Simias concolor* (fig. 3.3), suggest that they live in monogamous pairs, an unusual grouping for primates, formerly thought to occur only among gibbons, a few

Fig. 3.3. *Simias concolor*, the rare "simakobu" monkey of the Mentawei Islands, is suspected of being monogamous. (R. Tilson, personal communication.)

prosimians, the marmosets, sakis, night monkeys of South America, and, occasionally, man.

These exceptions aside, the important point is the recurrence of the same morphological and behavioral features among widely separated members of the subfamily Colobinae in Africa, India, and Southeast Asia. This complex of traits includes: a sacculated stomach; a flamboyant natal coat; the reduced importance of sexual swellings; a remarkable degree of situation-dependent tolerance between adult males; territoriality; long-distance male vocalizations; postnatal infant transfer; male take-overs; and infanticide (table 3.2). Though the possibility of environmental convergence cannot be ruled out, the case for phylogenetic inheritance of these traits among geographically disparate relatives is compelling and suggests that the ancestral colobine prior to the Africa-Asia split possessed similar life-ways or, at the very least, was predisposed to evolve them.[4] Far from being a recent, or "pathological" response to crowding, patterns of behavior such as male take-overs and infanticide may have been part of the colobine repertoire since Pliocene times. The finding that *Mesopithecus* was characterized by extremely dimorphic sexes is not out of line with the intense male-male competition that a social structure involving male take-overs implies (Trivers, 1972). Predispositions to the reproductive strategies of hanuman langurs may be quite ancient, dating back some ten million years or more.

The Flexible Hanuman

Presbytis entellus is the most ground-adapted of all the leaf-eaters. In forest areas, these langurs spend much of their days as well as their nights sitting in trees, but those living near open areas come down to the ground to feed and groom and may spend as much as 80 percent of the day there. On the ground, langurs rarely move too far from the safety of trees, and retreat up them whenever danger threatens. They also disappear into the shady treetops or, in arid areas, into rock shelters at midday to avoid the full heat of the

---

4. Consider, for example, natal coats. Though most colobines are born with flamboyant natal coats, the coloring of these coats differs remarkably between species. Hence the type of coloring may not be at issue. Rather, the particular pattern of social organization inherited by members of this subfamily may have created conditions under which postnatal infant transfer was particularly advantageous. As a consequence, over time natal coat color would be selected for greater and greater flamboyance in order to attract the attention of conspecific caretakers (Blaffer Hrdy, 1976).

TABLE 3.2. Basic features of colobines. + indicates reported for one or more study sites including zoo studies. – indicates not reported for any of several study sites. ? indicates insufficient information. Susp. indicates suspected.

Species and locations where most observations made	Male displays, long-distance vocalizations	Territoriality	Both multi- and one-male troops	Male take-overs	Infanti-cide	Female sexual swellings or reddening with estrus	Flamboyant natal coat	Postnatal infant transfer
Colobus polykomos West Africa	+	Susp.	+	?	?	–	+	?
Colobus guereza East Africa	+	+	+	+	+	–	+	+
Procolobus badius East Africa	–	–	Large multimale troops	–	–	+	– [black]	–
Procolobus verus Central Africa	?	?	?	?	?	+	–	?
Presbytis senex Sri Lanka	+	?	+	+	+	–	+	?
Presbytis johnii Western India	+	+	+	Susp.	Susp.	–	+	+
Presbytis entellus India and Sri Lanka	+	+	+	+	+	–	– [black]	+
Presbytis geei Assam	?	?	+	?	?	–	+ [black]	?

(continued)

TABLE 3.2. (continued)

Species and locations where most observations made	Male displays, long-distance vocalizations	Territoriality	Both multi- and one-male troops	Male takeovers	Infanticide	Female sexual swellings or reddening with estrus	Flamboyant natal coat	Postnatal infant transfer
Presbytis pileatus Assam	?	?	+	?	?	?	+	?
Presbytis obscurus Southeast Asia, Chicago Zoo	+	?	+	?	?	−	+	+
Presbytis cristata Malaysia	+	+	+	+	Susp.	−	+	+
Presbytis melalophos Southeast Asia	+	+	+	?	?	−	+	?
Presbytis potenziani Mentawei Islands	+	+	Monogamous pairs	?	Susp.	?	+	?
Pygathrix nemaeus Vietnam, San Diego Zoo	+	?	+	?	?	+	+	+
Nasalis larvatus Borneo	+	?	Mostly multimale	−	−	?	+ (face only)	?

Source: see table 3.1 and text.

sun. On cold winter mornings, langurs may descend onto rocky out-crops to sit together, their black faces tilted towards the sun, while warming up.

Far more omnivorous than the epithet "leaf-eater" suggests, langurs are able to digest large quantities of mature leaves, but they also feed on new leaves, seeds, nuts, fruit, grain, sap, insect pupae, and, according to one report, bird's eggs (Rahaman, 1973). This flexibility of diet, the capacity to survive in both wet and extremely dry conditions, and the ability to cover long distances on the ground—and occasionally to cover them quickly—combine to make the hanuman langur the most widespread primate species other than man on the Indian subcontinent.

Like another nearly omnipresent Indian monkey, the macaque (the rhesus macaque in the north, the bonnet macaque in the south), langurs in populated areas are concentrated near temples where they are fed by holymen and passersby, near bazaars in which they scavenge, and about crops, which they may occasionally raid. To date there is little information on the overlap and possible competition between langurs and macaques. Very generally, where macaques abound, there are no langurs; but one can find many exceptions. At Dharwar, for example, bonnets and langurs not only live in the same area but can be found feeding side by side in the same trees. In other areas, such as near Mussoorie in the Himalayas, a stricter separation between langurs and the sympatric rhesus macaques seems to prevail. Curiously, on the several occasions that isolated macaques have been known to join langur troops, the macaques have always managed to dominate their langur hosts (Dolhinow, 1972, for Kaukori; F. Singh, personal communication, for Sawai Madhapur Sanctuary). This finding is especially odd since experimental studies indicate that langurs are generally bolder than macaques (Singh and Manocha, 1966), and may reflect the greater commitment of macaques to rigidly hierarchical arrangements (chapter 6).

Despite the great variation between areas where langurs are found, similarities in adaptation, morphology, and social structure far outweigh any differences. Throughout their range, langurs are remarkably alike in appearance even though taxonomically they are considered different at a racial or subspecific level. The grey langurs of Sri Lanka (*Presbytis entellus thersites*) look and behave much like their continental counterparts, but carry their tails differently and have a peaked cap of fur on top of their heads.

Langurs in the south of the subcontinent (at the study site of Dharwar) demonstrate the same repertoire of behaviors as langurs in the north (at Jodhpur and Abu), but tail carriage among the southern langurs is different. At the southern site, langur tails form an S-shaped loop behind the animal while in the north, the tail loops back toward the head.

Based on this morpho-geographic difference in tail carriage, a distinction is frequently made in the behavioral literature between the "north" and "south" Indian langur; but because type of tail carriage does not seem to coincide with any particular behavioral pattern, this distinction between the two forms seems neither justifiable nor useful. From a behavioral point of view, the langurs at Jodhpur and Abu have more in common with langurs to the south at Dharwar than they do with their northern neighbors at Kaukori whose tails are carried with the same loop.

Some other differences found among the langurs living in the Himalayan regions, at the northern extreme of the species range, include the rarity of male whoop vocalizations and of female sexual solicitations. As with most colobines, langurs do not exhibit pronounced sexual swellings; instead, estrous females present their rumps to the male and frenetically shudder their heads. Even this visible signal of estrus is absent among the Himalayan population studied at Bhimtal (Vogel, 1971) and is rare at Melemchi (Bishop, 1975a). At all four Himalayan sites, including the sites of the Sugiyama (1975) and the Boggess-Curtin studies, the long-distance "whoop" vocalization by males is uncommon or absent altogether. Vogel has suggested that the omission of the long-distance calls from the langur vocal repertoire may be due to "deceptive echoes" in mountainous regions that prevent the whoop from fulfilling its original function of signaling location (1973b).[5]

5. Altitude alone cannot be at issue here. Though Vogel (1971) refers to Bhimtal as a "high altitude" study site, it is in fact at the same height as Abu (table 3.4.) where the repertoire of langur behaviors is identical to that of sea level populations. Two other Himalayan sites, Melemchi and Solu Khumbu, are among the highest locations where primates have ever been studied (the only higher study locations being for one gorilla study, one of Japanese macaques, and one of gelada baboons; Bishop, in press). Bishop (personal communication) attributes the omission of the whoop from the Himalayan langurs' repertoire to very low population densities. The langurs, according to her, quite simply have no one to call to.

Beginning with Phyllis Jay's pioneer research at the forest of Orcha and on the cultivated plains surrounding Kaukori, *Presbytis entellus* has been the object of eleven major studies, eight in India, one in Sri Lanka and two in Nepal. These eleven studies and the thirteen main sites where they were undertaken are listed in table 3.3. Locations of study sites are indicated in figure 3.4.

The langur studies differ substantially in both duration and in conditions of observation—mainly a function of how much time langurs spent in trees or in thick foliage, how habituated they were to human observers, and how many langurs at each study site were recognizable as individuals—as well as in ecological conditions. In table 3.4 basic environmental features of the study sites are summarized, including the extent to which humans affected the habitat in each location. Another important difference between sites is the degree to which langurs depend upon humans. At the game preserve of Orcha, or on the oak forested mountain slopes surrounding the isolated Himalayan village of Melemchi, langurs rarely encountered humans. Near the city of Jodhpur, however, on the edge of the Great Indian Desert, humans are an unescapable fact of the habitat, and probably essential to the dry-zone adaptation of langurs, who make use of man-made reservoirs and tanks and depend to some extent on crop-raiding and on provisioning by Hindus. Crop-raiding also accounted for a substantial portion of the langur diet at Kaukori and at the Himalayan site of Solu Khumbu (Curtin, 1975; Boggess, 1976).

As can be seen from the column on troop size in table 3.4, langurs typically travel in groups of 15 to 25 individuals, but in arid areas such as Jodhpur, Sariska, and Kaukori, troops are much larger, averaging between 35 and 64 individuals. This increased troop size for langurs living in dry, open areas has never been specifically investigated, but may reflect foraging conditions. In dry areas, animals cluster about water sources, provisioning centers, and garden spots. It is also possible that increased troop size provides additional warning and protection against predators in areas where there are few trees to flee to. Whatever the cause, Sugiyama was able to show that at Dharwar, in a south Indian region of alternating teak forest and fields, the size of langur troops living in open areas is slightly larger than the size of troops living in forests.

With the exception of Orcha, Kaukori, Melemchi, and Solu Khumbu, where langur population densities are exceedingly low, the mean areas of langur home ranges fall between 0.2 and 0.7 square kilometers. Roughly speaking then, 100 acres is about the space a troop of langurs needs. Because of variability in quality of

TABLE 3.3. Location and duration of *Presbytis entellus* studies between 1958 and 1975.

Observers	Years of study	Study sites	Observation hours	Number of months in field	Observation conditions
1. P. Jay	1958-59	Orcha Kaukori	Total, 850	12 4	Poor Excellent (individual recognition for one troop)
2. Y. Sugiyama S. Kawamura K. Yoshiba M. D. Parthasarathy D. Miyadi	1961-63	Dharwar	Not given	23	Good (individual recognition for many troops)
3. S. Ripley	1962-63 1968-71	Bundala and Polonnaruwa (Sri Lanka)	Ca. 3,000 2,135	Ca. 36	Good (individual recognition for several troops)
4. S. M. Mohnot	1967-	Jodhpur	Over 5,000	Permanent home	Excellent (individual recognition for several troops)
5. C. Vogel I. Weber H. Kruger	1968	Bhimtal and Sariska	1,071	3	Good (recognition of distinctive animals
6. S. Blaffer Hrdy D. B. Hrdy	1971-75	Mount Abu	1,503	11	Good (individual recognition for several troops)

7.	N. Bishop J. Bishop	1971-72	Melemchi (Nepal)	500 (contact hours)	11	Good (recognition of distinctive animals)
8.	H. Rahaman	1971	Gir Forest	Not given (few)	1	Poor (almost no individual recognition)
9.	R. Curtin J. Boggess	1972-74	Solu Khumbu (Nepal)	686 820	17	Good (individual recognition for several troops)
10.	Y. Sugiyama	1972-73	Simla	300	5	Good (little individual recognition)
11.	J. Oppenheimer	1971-72	Singur	5,000	20	Excellent (almost no individual recognition)

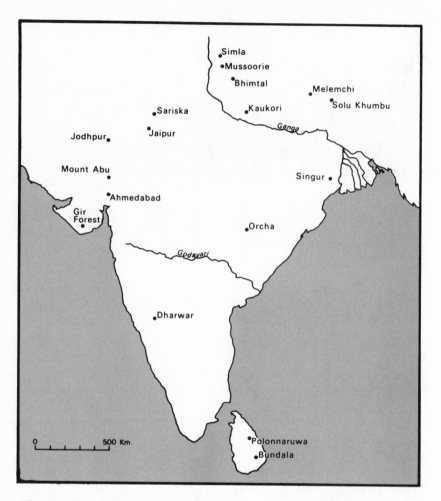

Fig. 3.4. Location of langur study sites.

ranges, and because there are so few data on the subject, it would be premature to speculate further on the relationship between troop size and home range size. What is evident, however, is that population densities are a good indicator of how often langur troops meet neighboring troops. As can be seen by comparing the column on population density from table 3.4 with the repertoire of behaviors for each study site in table 3.5., langurs at higher densities, or in areas where langurs are densely clustered about resources (as at Jodhpur), engage in frequent intertroop encounters, while those animals living at low densities almost never do. Encounters were

TABLE 3.4. Ecological features of major study sites where hanuman langurs have been observed for 100 hours or more.

Site and Source	Altitude (m)	Rainfall (cm) [% monsoon]	Population density (per km^2)	Habitat type	Human influence[a]	Mean troop size [range]	Home range (km^2)
MELEMCHI, NEPAL (Bishop, 1975b)	2,442-3,050	200 [measured only during monsoon]	Ca. 16	Himalayan temperate forest	Weak (some predators)	32 [1 troop]	2.1
BHIMTAL, KUMAON HILLS (Vogel, 1971; 1973)	900-1,500	260 [75%]	97	Upper monsoon forest	Strong (few predators)	23 [15-30]	0.2
SARISKA, RAJASTHAN (Vogel, 1971; 1973)	400	80 [80%]	104	Thorn forest	Weak (a number of predators)	64 [30-125]	.6
KAUKORI, U.P. (Jay, 1965)	122	76-127 [70-80%]	2.7	Dry scrub forest	Strong, much cultivation	54 [1 troop]	7.8
JODHPUR, RAJASTHAN (Mohnot, 1971a, b; 1974; personal communication)	241	28 [90%]	18	Arid open scrub	Strong, langurs in three types of habitats: concentrated around garden spots, in open scrub, near human habitations	35 [8-82]	Garden: .6-.96; Open rocky: .74-1.3

(continued)

TABLE 3.4. (continued)

Site and Source	Altitude (m)	Rainfall (cm) [% monsoon]	Population density (per km²)	Habitat type	Human influence[a]	Mean troop size [range]	Home range (km²)
ABU, RAJASTHAN	1,150-1,240	200 [90%]	50	Subtropical deciduous forest	Strong but little cultivation (few predators)	21 [15-30]	.3
GIR FOREST, GUJARAT (Rahaman, 1973; S. Berwick, personal communication)	226-648	108	—	Riverine deciduous forest	Fairly weak (many predators)	30 [16-48]	—
ORCHA, M.P. (Jay, 1965)	762	203 [75%]	2.7-6.0	Moist deciduous forest	Weak (game preserve with a number of predators)	22	3.9
DHARWAR, MYSORE (Sugiyama, 1964; Yoshiba, 1968)	50-760	76-127 [90%]	84-133	Dry deciduous forest	Fairly strong, includes both forest and fields (some predators)	14 [forest] 17 [open]	.2
POLONNARUWA, SRI LANKA (Ripley, 1965; 1967b; personal communication)	60	127-191 [75%]	100-200	Mixed evergreen and deciduous forest	Archaeological refuge (serious predation by dogs)	25 [12-42]	.25

SOLU KHUMBU, NEPAL (Curtin, 1975; Boggess, 1976)	2,591-3,505	244 [90%]	1	Mixed evergreen and deciduous forest and meadow	Moderate (some predators)	11	12.7
SIMLA, U.P. (Sugiyama, 1976)	2,000-3,000	164 [90%]	25	Himalayan moist temperature forest	Strong (few predators)	48 [8]	1.9
SINGUR, WEST BENGAL (Oppenheimer, in press and personal communication)	7	146 [90%]	5-20	Village	Strong (few predators)	13 [2 troops]	.04

a. Since the number of predators (such as tigers or panthers) is more or less inversely correlated with the presence of humans, to save space these two variables appear in the same column.

TABLE 3.5. Features of *Presbytis entellus* social behavior at major study sites where langurs have been observed for 100 hours or more. + indicates reported. − indicates none reported or else, in opinion of field worker, does not occur. ? indicates either no data available or else features not determined.

Study site	Predominant troop type [number of troops]	Intertroop encounters	Male take-overs	Adult male infanticide	Crop raids	Predictable female displacement hierarchy	Female sexual solicitation	Post-natal infant-sharing
Melemchi	Multimale [1]	Rare	−	−	+	?	−	+
Bhimtal	33% one-male [3]	Rare	?	−	?	?	−	?
Sariska	One-male [1]	Frequent	+	−	?	?	+	?
Kaukori	Multimale [1]	−	−	−	+	Exists, but poorly defined	+	+
Jodhpur	Mostly one-male [20]	Frequent	+	+	+	?	+	+
Abu	86% one-male [7 for 1971]	Frequent	+	+	−	+	+	+
Gir Forest	44% one-male [9]	?	?	?	?	?	+	?

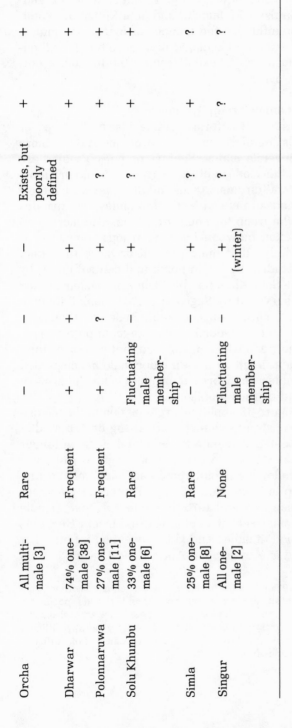

Orcha	All multi-male [3]	Rare	–	–	–	Exists, but poorly defined	+	+
Dharwar	74% one-male [38]	Frequent	+	+	+	–	+	+
Polonnaruwa	27% one-male [11]	Frequent	+	?	+	?	+	+
Solu Khumbu	33% one-male [6]	Rare	Fluctuating male membership	–	+	?	+	+
Simla	25% one-male [8]	Rare	–	–	+	?	+	?
Singur	All one-male [2]	None	Fluctuating male membership	–	+ (winter)	?	?	?

frequent at Dharwar, Polonnaruwa, Jodhpur, and Abu but exceedingly rare at Orcha, Kaukori, Melemchi, and Solu Khumbu.[6] Population density may also influence the number of extratroop males, the size of male bands, and the likelihood that these bands will encounter and be able to usurp bisexual troops. Infanticide is correlated with such take-overs.

Langur Troop Dynamics

Langur societies assume two forms: stable bisexual troops or nomadic all-male bands containing between three and sixty or more adult, subadult, and juvenile males. Solitary or paired males are sometimes spotted, but this mode of life is less common and probably temporary. Under special circumstances, adult females and their progeny travel for periods of time without adult males. Old females may sometimes leave the troop for hours at a time. The most commonly reported distinction between bisexual troops, however, is whether there is one fully adult male present or more than one. Whereas most troops studied by Jay in north and central India, by Curtin and Boggess at Solu Khumbu, by Bishop in Melemchi, by Vogel in Bhimtal, and surveyed by Sugiyama (1964) near Raipur in central India contained more than one adult male, at Dharwar, Jodhpur, Sariska, and Abu a predominantly one-male pattern prevailed. At Abu, only one troop was encountered that had more than one permanently resident adult male—an older male accompanied by two younger adult males who were subordinate to him in terms of both access to food and estrous females.

In a population of langurs, the political organization and composition of each troop is constantly changing. Drawing on the studies listed in table 3.3, one can construct a hypothetical model of langur social organization that takes into account observed variation in troop structure. The following hypothetical sequences of the life-cycle of a langur troop is illustrated by actual troop compositions recorded at Abu between 1971 and 1975. In the *initial stage,* a male from outside the troop usurps control of a bisexual troop. He evicts all other adult and subadult males and kills unweaned infants. For example, this occurred at Abu around April of 1972, when a male

6. According to Bishop (1975b), the low levels of any sort of aggressive interaction at the high altitude study site of Melemchi may be due to *both* wide spacing between troops and to the necessity of conserving energy in a harsh environment where occasional snow and freezing temperatures make life generally difficult. Bhimtal and Sariska have been omitted from this discussion because of the difficulty of assessing population densities from short studies.

(designated Shifty Leftless) took over a troop known as the Bazaar troop, ousting three adult males and one juvenile male in the process. Subsequently, a second juvenile male also left the troop. Five infants disappeared at the time of this change and were presumed to have been killed. These events were repeated twice more in the Bazaar troop before this study ended, as Shifty was replaced by a male called Mug and then Mug himself was replaced by Righty Ear. These changes in the Bazaar troop between August 1971 and June 1975 are summarized in the troop compositions provided in table 3.6.

Following the first Bazaar troop take-over, by March of 1973, four new infants had been born in Shifty Leftless' troop; one subadult female had become pregnant, and three juvenile females had begun to cycle. With such increases in the ranks of child-bearing females, the Bazaar troop entered a *growth stage*. The remarkably fertile School troop perhaps provides the best illustration of this stage. In June of 1971 the School troop consisted of a single adult male and eight females, every one of which was associated with an infant between the age of eight months and one year. Again in March of 1973, four females carried newborn infants while the remaining three were in the last weeks of pregnancy; the eighth female had disappeared. Nevertheless, in the four years of this study, the number of animals in the School troop never exceeded twenty-one individuals, even though the same alpha male (Harelip) retained control for the entire four years. One reason for this constancy despite the birth of several crops of infants was that many infants born were males who disappeared from the troop prior to reaching adulthood. Between 1971 and 1973, another Abu troop (labeled the I.P.S. troop) grew from twenty-four individuals to over forty. By 1975, however, this burgeoning troop had split into two parts, each captained by a new adult male. What became of the old leader was not known.

In some cases, the growth process may continue through a *mature stage*, when after five years or more all age grades are represented. The Bazaar troop in 1971 and the Toad Rock troop in 1973 (see table 3.7) may have been on the verge of such a pattern, but in both cases full maturation into an age-graded pattern (described by Eisenberg et al., 1972) was interrupted by male take-overs, returning the troop to the initial stage of the cycle.

Take-overs resulting in one-male troops are well documented. The unresolved question remains: how do multimale troops came into being? Because all of the long-term langur studies (at Dharwar,

TABLE 3.6. Changes in the composition of Bazaar troop between August 1971 and June 1975.[a]

Date	Adult male	Adult female	Infant-1	Infant-2	Juvenile	Subadult
August 1971 (total = 22)	Alpha	1.	1. (born late July)	Junebug (f; 14 months)	Breva (f)	Elfin (f)
	#2	2.	2. (born late July)	2. Male (15 months)	Guaca (f)	Nonyno (f)
	#3	3.	..... 3. (born early August)		Collias (m; 16 months)	
		4.			4. Male	
		5.				
		6.				
		7.				
		Quebrado				
June 1972 (total = 14)	Shifty	Nonyno			Collias	Elfin
	Leftless	Overcast			Junebug	Guaca
		Kasturbia				Breva
		Earthmoll				
		Pout				
		Quebrado				
		Wolf				
		Short				

March 1973 (total = 17)	Shifty	Elfin		Breva
	Leftless	Nonyno		Junebug
				Guaca

Overcast
Kasturbia ...... Nuka (f; 1 month)
Earthmoll ...... Quilt (m; 4-5 months)
Pout .... Astra (f; 3 months)
Quebrado
Wolf .... Bump (m; 4-5 months)
Short

| January 1974 (total = 19) | Shifty | Elfin ....Elf (m; 3-4 months) | Junebug |
| | Leftless | | |

Nonyno .... Nonet (f; 10 months)
Kasturbia .... Nuka
Breva
Earthmoll .... Quilt
Pout .... Astra
Guaca .... Guat (f; born Jan. 10)
Overcast
Wolf .... (Bump 14 months; died Dec. 20)
Quebrado
Short

(continued)

TABLE 3.6. Changes in the composition of Bazaar troop between August 1971 and June 1975. (continued)

Date	Adult male	Adult female	Infant-1	Infant-2	Juvenile	Subadult
April 1975 (total = 24)	Mug[b]	Elfin .... newborn				Nullipara #2[c]
		Nony				Nullipara #1[c]
		Breva .... Brief (f; 12 months)				
		Pout .... 2 months				
		Junebug .... newborn				
		Guaca .... Guat				
		Kasturbia ...... newborn				
		Overcast				
		Earthmoll ...... newborn				
		Short .... newborn				
		Quebrado				
		Wolf .... newborn[d]				

a. M indicates male; f indicates female. Dots connect females with infants presumed to be their offspring.

b. In April 1975 Righty Ear replaces Mug; two newborns die.

c. Of three young females (Nonet, Nuka, and Astra) present in 1974, only two remained; it was not possible to definitely identify which two.

d. This infant died in an accident October 19, 1975 (personal communication, James Malcolm).

TABLE 3.7. Changes in the composition of Toad Rock troop between July 1971 and June 1975.[a]

Date	Adult male	Adult female	Infant-1	Infant-2	Juvenile	Subadult
July 1971 (estimate)	Splitear	8	1?	3 females	3 males	2 males 2 females
July 1972 (total = 23 or 24)[b]	Splitear	Pawlet		....	Hauncha (f; 18 months)	Beny (m)
		Mopsa	Brumio (m; 4-5 months)	1. Male	Pandy (f; 15 months)	No. 2 (m)
		C.E.		2. Male	Cast-Eye (m; 30 months)	Scrapetail (f)
		P-M T.T. Mole		3. Male		
		I.E.		4. Male	Handy (f; 15 months) 5. Male (36 months) 6. Male (36 months) 7. Female (15 months)	
		Old female who disappears				

(continued)

TABLE 3.7. Changes in the composition of Toad Rock troop between July 1971 and June 1975. [a]   (continued)

Date	Adult male	Adult female	Infant-1	Infant-2	Juvenile	Subadult
February 1973 (total = 23)[b]	Splitear	Mopsa Mole Pawlet P-M C.E. I.E. T.T. Scrapetail	Niza (f; 1 month) N-M (f; 1 month)	Brumio	Hauncha Handy Pandy No. 4b Cast-Eye No. 3 Blind-ear 8. Male 9. Male 10. Male	No. 2
January 1974 (total = 18)	Toad	Mole Mopsa I.E. T.T. Scrapetail C.E. P-M Pawlet	Moli (f; born Jan. 14) Brujo (m; born Jan. 9)	N-M Vert (f; 6 months) Niza	Hauncha Handy Pandy No. 4	

April 1975　　Toad?　　Scrapetail　　　　　　　　　　　　　　　N-M　　Hauncha
(total = 19)　　　　　　Pawlet　　　　　　　　　　　　　　　　Niza　　Pandy
　　　　　　　　　　　　Mole　　　　　　　　　　　　　　　　　Moli　　Handy
　　　　　　　　　　　　P-M .... P-M's　　　　　　　　　Vert　　No. 4
　　　　　　　　　　　　　　　　　　(f; 6 months)
　　　　　　　　　　　　T.T.
　　　　　　　　　　　　I.E. .... I.E.'s
　　　　　　　　　　　　　　　　　　(f; 5 months)
　　　　　　　　　　　　C.E. .... C.E.'s
　　　　　　　　　　　　　　　　　　(f; 5 months)

a. M indicates male; f indicates female. Dots connect females with infants presumed to be their offspring.

b. A fourth subadult female was definitely present in the troop in December 1973. This female was not, however, included in the counts for previous years. Two possibilities come to mind: either this young female joined the Toad Rock troop sometime between September and December 1973 or else, as seems more probable, I made an error in earlier years possibly counting Pandy and No. 4 as the same animal.

Jodhpur, and Abu) are in areas where one-male troops predominate, the process by which multimale troops are formed has never been described. There are at least two possibilities that would account for the presence of more than one adult male in a troop: multiple male invasions, and a gradual process in which a single adult male retains control of a troop long enough (five to seven years) for his own sons, who are tolerated by him, to mature. This second hypothesis provides an attractive explanation for the very stable relations which were observed between males at Orcha, Kaukori, Melemchi, and elsewhere. Provided that they are not driven out, males growing up in a troop could possibly ease into possession of it, as may have occurred in a peaceful transition that Jay observed at Kaukori (1965). Nevertheless, there are no published accounts of adult males tolerating young males known to be their sons. Furthermore, at Abu, of the many males born in the five troops that we followed over a four-year period, not one remained in his natal troop long enough to reach adulthood. The tension evident between these young males and the troop leader, together with actual observations of young males being driven out by the alpha male among langurs at Jodhpur (Mohnot, 1974)—even in cases where the alpha male was almost certainly the father of these youngsters—indicates that sons are not necessarily tolerated.

Regardless of how multimale troops came into being, there is a striking (but not perfect) correlation between low population densities and the existence of multimale troops. Very probably, the number of adult males present in a langur troop reflect population conditions and historical factors rather than innate differences between langurs in different areas. To the extent that the numbers of extratroop males and the frequency of contact is correlated with the likelihood that such males will take over a troop, environmental factors such as high population density contribute to the prevalence of one-male harems. Certainly with large numbers of extratroop males, as is the case at Dharwar, Jodhpur, Abu, and Sariska, a male take-over appears to be a far more likely outcome than maturation of the troop and an intratroop transfer of control.

Because of recurrent fluctuations, it is important to define carefully terminology applied to langur groups. As used in this study, *troop* refers to any stable association of females and their offspring living in a more or less fixed geographic location or *home range*. The term "troop," and the name given that troop (based on some geographic feature of their home range), applies regardless of which or how many adult males are present. It appears that such troops are closed and exceedingly difficult for any alien female to join. Such an occurrence would be rare, if it occurs at all. Males on the other hand regularly enter bisexual troops despite the resistance of females and especially of other male troop members. Troop histories from Dharwar and Abu indicate that under current conditions at these sites, new males actually take over troops every two to three years.

Following male take-overs, adult females (usually mothers) have on occasion disappeared from their troops or left temporarily. Whereas temporarily absent females are known to have traveled with males ousted from their troop, with other mothers from the troop, or on their own, the destination of females who never returned to their troop after a take-over is unknown. Death and solitude are two possible fates for such missing females, joining another troop a third, but so far undocumented, possibility. Among the related species *Presbytis senex*, however, immature females as well as males are sometimes driven out by usurping males and may remain in the company of male bands in what Rudran refers to as "predominantly male troops" (1973b).

It is often useful to specify whether at a given time a bisexual troop is a *multimale troop* containing more than one resident adult male or a *one-male troop* or *harem* which contains only one full adult male. In the case of a troop with only one adult male, or in the case of a multimale troop or male band in which one adult male is able to displace all other animals in the group, that male is referred to as the *leader* or *alpha male*. *Alpha female* refers to the female in the troop able to displace all other females in the troop. All such hierarchical terms refer to specific points in time. Unless otherwise specified, the terms *dominant* and *subordinate* refer only to the ability of one animal to displace another for access to food, or a spatial position.

Extratroop males refers to any male living outside of a bisexual troop. Such males living in either temporary or semipermanent associations are referred to as *all-male bands*. Because of their nomadic way of life and fluctuating numbers, names given to male bands only apply to specific periods of time; such bands lack the continuity of bisexual troops, which remain in the same location generation after generation. The term *invader* designates any extratroop male who invades a bisexual troop, and is used even if that invading male subsequently becomes the resident alpha male of the troop. The term *alien* refers to any animal belonging to another group, whether it is another troop or a male band.

Armed with these definitions, we can consider what the ramifications of langur demography are for the genetic and social relationships of the animals involved.

Genetic Consequences of Langur Troop Structure

In the course of his career, a male langur may belong to a number of different groups, including his natal troop, bisexual troops he subsequently invades, and male bands. A female, on the other hand, remains perpetually in the same troop for some twenty years or more. The stable social unit is composed of overlapping generations of matrilineal relatives (color fig. 1). As a consequence of this system, females and offspring belonging to the same troop will be closely related. By contrast, the degree of relatedness between females and incoming males is much lower except in the case of an ousted male returning to his natal troop. Unfortunately, the likelihood of a male ever returning, either by preference, chance, or opportunity, is not yet known. Nor do we have data concerning the background of inbreeding present in any natural population of langurs. Only the most preliminary stages of such research have been undertaken (D. B. Hrdy, Barnicot, and Alper, 1975).

It is assumed here that by and large the degree of relatedness between an incoming langur male and the females in his new troop is low, and that a single adult male in a troop, will be the progenitor of most, if not all, infants born in that troop during his tenure. Nevertheless, the first assumption is an oversimplification, while the second discounts possible genetic input from copulations between troop females and extratroop males. For the purposes of estimating degrees of relatedness between given individuals in a langur troop, a third assumption is made, namely, that female membership is stable over long periods of time and results in an average degree of relatedness among adult females of 1/8, that is, about as close as first cousins. Table 3.8 gives hypothetical minimum degrees of relatedness for individuals in a small troop of langurs. These coefficients of relatedness should be regarded as only very rough approximations based on assumptions that will undoubtedly be refined as more information becomes available (see also appendix 3).

From the assumptions made, and from patterns of behavior observed after take-overs (that is, the killing of unweaned infants in the troop and the return of their mothers to sexual receptivity), it follows that the degree of relatedness between an incoming male and infants born one gestation period later will be at least one-half. Females in the troop will be more or less equally related to the in-

TABLE 3.8. Hypothetical minimum degrees of relatedness (that is, the likelihood of having genes shared by common descent) for individuals in an average langur troop of 25 animals.[a]

Category	Adult female	Incoming male	Offspring of previous male	First cohort conceived after take-over
Adult female	1/8 on average	Assumed 0	1/2 - 1/16	1/2 - 1/16
Incoming male		—	0	1/2
Offspring of previous male				1/2 - 9/32[b]	1/4 - 1/32
First cohort conceived after take-over				9/32

a. This table is based on the assumption that female membership in the troop is stable over time and that all adult females in the troop are related by at least one-eighth. Also an average male tenure of three years is assumed. The possibility that entering males are significantly related to females in the troop is ignored. Hence, these are minimum estimates because they do not take into account the background of inbreeding, the magnitude of which is as yet undetermined for langurs living on Mount Abu. For a derivation of these values and a discussion of factors that might alter them, see appendix 3.

b. This category may comprise cohorts of infants born over a three-year period and hence some full siblings, born to the same mother as well as same father, will be included.

fants killed as to the new infants sired by the invader. There is a potential asymmetry, however, in the degrees of relatedness between cohorts (for example, between juveniles and infants), which will depend on the length of a given male's tenure. Whereas at the outset of a male reign the degree of relatedness between females sets the maximum limit for the degree of relatedness between members of different cohorts, later on different cohorts will be related through both parents. When a new male has recently entered the troop and sired offspring, older infants and juveniles are more closely related on average to subadult and adult animals in the troop than to any subsequently born infant. Because of the possibility of an animal either augmenting or decreasing his or her inclusive fitness by cooperating with or exploiting another individual, degrees of relatedness are highly relevant to analyses of cooperative, competitive, and aggressive interactions among langurs.

These genetic relationships are the outcome of social patterns that almost certainly derive from a basic colobine adaptation. Harems with multimale potential but with the additional alternate possibility of male take-over—usually accompanied by infanticide—recur again and again among widely separated members of the subfamily Colobinae. Recourse to phylogeny, however, begs the larger question: why did males seek possession of females in the first place? A whole theory has grown up to explain why it is among so many different creatures that males compete for females.

As set out by Darwin, and subsequently refined by R. A. Fisher, George Williams, Robert Trivers, and others, natural selection is the process by which genes become disproportionately represented in a population through the differential reproductive success of some of its members. When differential success is the consequence of competition between members of one sex for access to the other sex, Darwin (1859) suggested that the term *sexual selection* rather than natural selection would be more precise; for the unsuccessful competitor in this struggle, the outcome is not death, but few or no offspring. In a recent refinement of the concept of sexual selection, Trivers (1972) has pointed out that the key variable determining the extent of intrasexual competition is the relative amount of parental investment by each sex. Defining parental investments as "any investment by the parent in an individual offspring that increases the offspring's chance of surviving (and hence reproductive success) at the cost of the parent's ability to invest in other offspring." Trivers argues that whichever sex's typical parental investment is greater will become the limiting resource for the other sex. Individuals of the sex investing less in an offspring will compete among themselves for the sex investing more. Because in almost all mammalian species female investment in offspring (which includes physiologically costly gestation and lactation periods) is greater than that of males, males compete for females.

Langurs are no exception. Aside from insemination, the main contribution made by a langur male to his progeny is to prevent another langur male from killing them. Interestingly, a male reaches the limit to the number of offspring he can successfully protect long before he reaches the limit to the number of offspring he is capable of fathering. As I was to learn at Abu, this discrepancy creates a basic dilemma in the life of a langur male.

Females on the other hand have a definite physiological ceiling on the number of offspring they can produce and rear in a lifetime. Maximizing reproductive strategies for a female entail ensuring that any offspring she produces—already costly by the time they are born—survive, and minimizing the interval between births so that she comes as close as possible to the childbearing limit of which she is capable. A female's optimum strategy reflects some balance between these two policies. Throughout this book, animals will be described as strategists, and it should be clear at the outset that no conscious calculations are necessarily implied. Rather, animals predisposed to respond to a given situation in particularly advantageous ways in the past contributed differentially to the next generation's gene pool. As used here, strategies are evolved, not invented.

Competition among males for females is basic to the colobine pattern of male take-overs and infanticide. Even more basic, however, may be the specifically female adaptation to the loss of an infant which enables males to profit from killing infants in the first place.

Reproductive Features of the Limiting Sex

The capacity of a female to reproduce, and especially how often she is able to reproduce, imposes a ceiling on the reproductive success of both sexes. Among langurs, a long period of maternal investment in each infant is combined with an unusually flexible breeding physiology.

Langur females begin to cycle in their third year of life, conceiving for the first time in the second half of that year. Under normal circumstances, langurs exhibit a typically primate menstrual cycle: estrous behavior around the time of ovulation, followed by menstruation—a skimpy flow which is not always detectable—and then estrous behavior again roughly 13 to 21 days after the menses, about one month after the previous estrus. The average number of days between normal cycles at Abu (based on ten cycles) ranged between 24 and 34 days, with the average being 28 days; this figure is close to the 26.8±1 days that Ramaswami and David (1969) found for 76 cycles among individually caged langur females studied for three and one-half years at Jaipur, Rajasthan.

There are no obvious morphological changes to signal sexual receptivity. Instead, the langur female solicits a male by presenting to him and frantically shuddering her head (fig. 3.5). Because in the field it is not possible for humans to distinguish between females who are ovulating and females who are not, the term *estrus* as used

Fig. 3.5. A female langur solicits a male by presenting her rump and frenetically shuddering her head.

here can only refer to behavior, not to any physiological state (Loy, 1970). If in retrospect it is obvious that the female could not have ovulated at the time she exhibited estrous behavior (that is, if she gave birth a few weeks later), than the term *pseudoestrus* is used to indicate that she definitely was not ovulating.

Based on field observations by Sugiyama in South India and on studies of caged langurs at the National Center for Primate Biology at Davis (L. J. Neurater, personal communication), gestation periods for hanuman langurs are assumed to be 200 ± 10 days or between six and seven months. In this study, approximate dates of conception were calculated by counting backwards six and one-half months, or to an estrous period observed between six to seven months before the infant was born. In cases where five or fewer months elapsed between the last observed estrous behavior and birth of a live infant, sexual solicitations were attributed to post-conception pseudoestrous.

Under some circumstances, langur females are continuously sexually receptive, a pattern previously thought to occur only among human females.[7] Social rather than purely physiological factors may affect the occurrence of estrus in important ways (Blaffer Hrdy, 1975). For example, it appears that when alien males enter a langur troop, some females exhibit estrous behavior day after day irrespective of their menstrual schedule, and may exhibit estrus and copulate even if they are already pregnant. In most cases, mothers exhibiting continuous estrus were pregnant or else suckling an infant (in each case an older infant). Such a switch to nearly continuous estrus may occur if the alpha male is replaced by a new male, or if invaders join the troop in addition to the old alpha. If a female's infant is killed by invaders, she enters sexual receptivity within hours or days, and from that point on cycles normally. Under the special circumstances of a troop take-over, immature females who have not yet begun cycling sometimes also exhibited estrous behavior, though males solicited by them were never observed to mount.

The onset of estrus may be influenced by social factors even under normal conditions. As has been reported for several primate species, including hamadryas baboons (Kummer, 1968), purple-faced leaf monkeys (Rudran, 1973a), and humans (McClintock, 1971), it is possible that langur females also tend to synchronize their cycles with female troopmates, though the sample size in this study was too small to absolutely rule out the possibility that the apparent synchrony was coincidental.[8] Recent experimental work by McClintock (1974) indicates that pheromones are the signaling agent for synchrony in rats, and probably for primates as well.

Though social factors appear to have an overriding influence in sparking sexual activity among langurs, the natural environment

7. There is at least one study indicating that human females may be more likely to copulate at midcycle when they are ovulating (Udry and Morris, 1968). Because langurs under some conditions opportunistically exhibit continuous receptivity and because some humans exhibit a midcycle peak in sexual behavior, it is difficult to draw a firm line between humans and other primates in this respect.

8. Even more tentative (and more intriguing) is the possibility that such synchrony may induce shorter cycle length among some females. The two shortest cycles recorded at Abu (Blaffer Hrdy, 1975) occurred when these females were in synchrony with others. Because of the implications of such short cycles for reproductive competition between females, it would be of great interest to have more data on this phenomenon.

also plays a significant role in the scheduling of langur reproduction. Several authors have noted seasonal peaks in the number of infants born in langur troops at various sites (Prakash, 1962; Jay, 1965; and Sugiyama et al., 1965a). This was also true at Abu, where newborn infants are almost certain to be present in some troops between the months of December and March. In addition, a number of births can be seen during the monsoon months of July and August, when much breeding behavior is also going on. Nevertheless, unlike some primates (such as the rhesus macaques of Cayo Santiago Island), at no langur study site (with the possible exception of the Himalayas) have births been reported to be strictly seasonal; typically, some births occur throughout the year.

Since the maximum parental investment required from a female may be while she is pregnant and lactating, it is curious that females should tend to carry their infants and to give birth during the driest part of the year when food is relatively scarce. For those females giving birth in midwinter, only conception and the earliest months of gestation occur during the wet season. It may be significant, however, that an infant born in December or January would be taking some solid food by the time of the next monsoon. Maximum availability of nutritious food would coincide then with an exceedingly vulnerable transition point in the life of the weaning infant. This same sequence of events has been reported in a study of female reproductive cycles among purple-faced leaf monkeys in Sri Lanka (Rudran, 1973b; also see Petter-Rousseaux, 1968).

In addition to a flexible breeding system, a second feature of langur reproduction—exceptionally long birth intervals—may predispose them to the special patterns of behavior they have evolved. Unlike other Old World monkeys who produce an infant annually (for example, the rhesus macaque), langurs are more nearly bi-annual breeders with birth intervals lasting anywhere from 15 to 30 months; during much of this time, a mother is nursing her previous infant.

The age at which langur mothers first discourage suckling varies widely between study sites and between individuals at the same site. Whereas at Orcha and Kaukori weaning took place at 15 months, at Dharwar it did not occur until 20 months (Yoshiba, 1968). At Abu, under certain conditions a mother may discourage her infant from nursing as early as 6 months (see chapter 8). But under normal circumstances, the onset of weaning is later, about 8 to 10 months, and lasts until ages 13 to 20 months. Both social conditions (for example, the presence of an infanticidal male) and environmental factors such as food availability may influence the timing, but to date no specific study of the factors affecting weaning has been made.

The birth interval is the sum of gestation time plus lactation period plus whatever lag may occur between the end of lactation and the next conception. Hence, in the case of the langurs of Abu, birth intervals would normally be between 20 and 30 months. Similar two-to-three-year birth intervals have also been reported by Sugiyama at Dharwar (1967). In some cases, females have conceived while still lactating, and in such instances birth intervals would be shorter than those suggested here. For example, one middle-aged Toad Rock troop female, "P-M," gave birth to two infants only 15 months apart. On the other hand, in harsh habitats the intervals may be much longer; under the near-desert conditions at Jodhpur the period between the end of lactation and the next birth can be as long as 27 months.

To sum up, "Being similar to man, the (she-) ape has a menstrual cycle governed by the moon" (Saint Hildegard of Bingen, about hamadryas baboons, in the twelfth century). Every 28 days a cycling langur exhibits estrous behavior. If she conceives, gestation lasts about six and one-half months. A female may continue to exhibit estrous behavior during pregnancy, and may be especially prone to do so if a male from outside the troop enters it. During such invasions, there may be a breakdown of cyclicity such that sexual receptivity becomes essentially continuous. After an infant is born, the mother may continue to nurse it for as long as twenty months, but usually for some shorter period. In general, a female in her prime will give birth every two to three years. Though some seasonality of births is apparent, a female who does not conceive during periods of maximum sexual activity or who loses her infant may conceive at another point during the year.

4 / Mount Abu and Its Langurs

The forested hillsides of Mount Abu rise steeply from the parched and overgrazed plains of southwestern Rajasthan. On clear days other members of the Aravalli chain, of which Mount Abu is a detached member, can be seen in the distance. Isolated stands of trees border cultivated fields and homesteads; camels and other livestock cluster about them for shade. The boulder-strewn, uncultivable hillsides support the only remaining forests, remnants of a time not very long ago when all of India was covered with jungle.

As the road winds up the mountainside from Abu station on the plains, vegetation grades rapidly from arid-zone scrub to thick deciduous forest. Interspersed through the forest is an occasional expansive fig or mango tree that retains its leaves year round. During the wet season, torrents of water cascade down rocky river-beds and the thick undergrowth becomes nearly impenetrable (fig. 4.1). In the dry season streams dry up, as do the leaves of most trees and shrubs, leaving a wasteland of dendritic brown skeletons (fig. 4.2); the few remaining grassy patches are scorched and flattened.

With the beginning of the forest it is possible to see langurs and to hear the haunting calls of peafowl. Langurs share this forest refuge with jungle fowl, partridges, the omnipresent bulbuls, pigs, hares, porcupines, palm civets, sloth bears, panthers, and the shy and

Fig. 4.1. During the monsoon months from mid-June through August, the hillsides are enveloped in green jungle.

shaggy sambar deer. Competing with these creatures for their forest are a number of tribal people who make their living primarily as woodcutters and herders. Though scarcer than on the plains, cattle and goats amble about these hillsides, occasionally browsing on the same food sources as langurs.

The main road up the mountain terminates 1,170 meters above sea level at the town of Abu, a hill station beside a year-round lake. Nearly 8,000 people are permanent residents, but thousands of Indians come yearly as pilgrims or tourists to this ancient religious center and its surrounding forests. The twelfth-century marble temple at Dilwara, the site at Achaleshwar (where Shiva's big toe is thought to be enshrined), and the various *gophas* or sacred caves that have been weathered out of the granite outcroppings on the hillsides are the primary attractions for the transient population (fig. 4.3). In addition, Abu is a favorite resort for honeymooners from the neighboring and comparatively wealthy state of Gujarat. The tourist trade together with large Indian army and Indian Police Service installations are the basis for the town's economy. There is little commercial cultivation at Abu, and no industry. Because of the

Fig. 4.2. By the end of winter, many trees have lost their leaves, and the hillsides take on a parched appearance. (Photo taken from halfway up the mountain in March 1973; desiccation at the top of the mountain is less pronounced.)

burgeoning number of visitors, the town is prosperous and undergoing rapid expansion and modernization.

I journeyed to India and to this hilltop town five times, and for more than 1,500 hours observed langurs at close quarters (table 4.1). Whereas the first two studies roughly coincided with the rainy months, the 1973-74 study was undertaken during the cold months of the winter dry season. The 1973 period spanned the dry season transition from cold to warmer weather, and finally, the fifth study period in 1975 coincided with the hot tailend of the most severe drought in Abu's recent history. During the 1971 study period, and for three weeks each in 1972, 1973-74, and 1975, I was assisted by Daniel Hrdy. Brief comparative observations were also made of langurs living near the city of Jaipur, in the city of Ahmedabad, at Jodhpur, Gir Forest, Dharwar, and the Himalayan hillstation of Mussoorie.

The studies at Abu consisted of brief roadside surveys made in a taxi and intensive observations of seven bisexual troops chosen for their accessibility. All behavioral observations were made on foot and at close range. Binoculars were rarely needed except to watch monkeys in the trees or on rooftops or to keep track of the movement

Fig. 4.3. "Toad Rock" is one of a number of oddly weathered granite boulders standing like prototypes for the work of Henry Moore about the craggy hillsides of Abu. This particular rock is a main tourist attraction and is also at the center of the Toad Rock troop's home range. In the background can be seen the former summer palace of the Majarajah of Jaipur, now a semiabandoned hotel where we set up our headquarters.

of more than one troop at a time. The seven langur troops varied in their contact with humans. Two troops, the Chippaberi bus stop troop (half way between the hilltop town and the railway depot on the plains below), and a second troop, the Bazaar troop, which moved within the confines of the main part of town, lived in close and commensal association with man, obtaining a substantial portion of their daily food either directly or indirectly from people. The langurs at Chippaberi were fed peanuts, chickpeas, and

TABLE 4.1. Study periods at Mount Abu.

Period	Hours of observation	Season
June 28–September 13, 1971	342.3	Monsoon
June 19–September 12, 1972	377	Monsoon
February 11–March 26, 1973	286	Dry
December 20, 1973–January 31, 1974	282.2	Dry
April 10–June 20, 1975	215.7	Dry

chappati (pancake style Indian bread) by travelers whenever a bus stopped to cool its engine, usually many times a day. Similarly, about once a day the Bazaar troop entered the shopkeeping district of town where it scavenged (fig. 4.4). Individuals in either of these troops could be approached within one meter. The other troops studied (Arbuda Devi; I.P.S.; Toad Rock; Hillside; and School troops) lived on the outskirts of town and were fed less frequently; they nevertheless quickly learned to tolerate the close proximity (within three to six meters) of a passive human observer.

Male bands encountered by these seven troops were also studied, but were more difficult to observe because of their initial shyness and their habit of suddenly moving off through the forest, often covering long distances in short periods of time. The number of hours that each of these seven troops and eight male bands was studied is summarized in table 4.2.

By keeping track of how long observations of each troop lasted, roughly accurate frequencies for the occurrences of certain events could be computed—for example, the frequency of encounters between neighboring troops, the amount of time invading males spent in a given troop, and so on. A more rigorous focal-animal sample technique (described in chapter 7) was used to collect information on specific problems, usually having to do with the care and transfer of infants.

Feeding by pilgrims and local people is a routine event in the lives of the langurs of Abu. In addition, there were four contexts in which I fed langurs with "gram" (chickpeas) or groundnuts. Langurs were fed to lure them close enough to squirt wth identifying stains, or to make quick counts of a troop not being specifically observed on that day; neither of these two types of provisioning directly involved behavioral observations. This was not the case for "food tests," which were performed by placing food between two or more individuals in order to establish which had priority of access, or, in a few instances, between two troops to determine if one troop could displace another over food. Such experiments were undertaken only *after* an encounter between two troops had already begun. Despite its usefulness as a research tool, there is little doubt that provisioning increases aggression between individuals.

This research was originally undertaken to investigate the phenomenon of infant-killing by adult male langurs and to ascertain why it occurred. Later, the study was expanded to include the reproductive strategies of both sexes. Because of these interests, attention was focused on rare rather than daily events. The pre-

Fig. 4.4. Langurs scavenge in the bazaar, alert to the possibility of un-guarded delicacies.

TABLE 4.2. The number of hours that each of seven troops and various male bands were observed at Mount Abu during five study periods.

Group	1971	1972	1973	1973-74	1975	Totals
Troop						
Chippaberi	77	2	0.5	—	4.5	84
Arbuda Devi	17	—	—	—	—	17
Bazaar	106	29	88	84.5	64.5	372
I.P.S.	48	16	6	7.5	15[c]	92.5
Toad Rock	18	101	34	133	90	376
Hillside	42	183	156[a]	16.5	25	422.5
School	22	41	1.5	24	11	99.5
Total:	330	372	286	265.5	210	1463.5
Male band						
Band of ca. 60 males, near Chippaberi	4	—	—	—	—	4
Waterhouse gang	—	—	—	5.3	—	5.3
M-1 (near Chippaberi)	2	—	—	—	—	2
M-2 (near Chippaberi)	1.3	—	—	—	—	1.3
Skulking duo	5	5	—	—	—	10
Righty's group	—	—	b	.3	—	.3
Splitear's band	—	—	—	11.1	—	11.1
Youngsters	—	—	—	—	5.7	5.7
Total:	12.3	5	b	16.7	5.7	39.7

a. Includes observations of five males who temporarily traveled with Hillside troop.

b. Observations of Righty Ear in 1973 are counted among the hours Hillside troop was observed.

c. By 1975, I.P.S. troop had split into two troops; the larger segment was watched 12 hours, the smaller 3.

vailing bias of this study was that those spheres of action where the mechanisms for maximizing reproductive success would be detectable to a field observer were sought out. That is, if an estrous female or a newborn infant or invading males were present in a troop, the observation time allocated to that troop was greater than it would have been if life in the troop was proceeding "as usual."

For a balanced portrayal of daily activity in a langur troop, the detailed records made by the Japanese team at Dharwar, especially Yoshiba's 1967 account and the numerous reports by Sugiyama, are superb and will be referred to throughout this book.

The unavoidable gaps of from six to fifteen months between study periods and of several days within study periods, when information from one troop was sacrificed in order to follow another, are serious disadvantages to this study because they interrupt the continuity of records for any individual or troop. This disadvantage was only partially offset by the advantage of working in an area where langurs live in association with some 8,000 scattered people. Unusual occurrences in the lives of the langurs were often witnessed by humans. Where reports from reliable informants coincided with what I knew about troop locations and changes in troop compositions, these reports were accepted by me as facts and are cited in this research as personal communications. On balance, the policy of intermittent observations over a number of years did have the advantage of putting individual reproductive strategies in long-term perspective.

The human population at Abu affected this study in several important ways. In areas around temples and human habitations, the langurs have been provisioned for many centuries. This conditioning of langurs to humans as a source of food makes Abu a poor place to undertake a naturalistic study of langur ecology but an excellent location for a series of yearly behavioral observations, since the animals are already habituated. In addition, observations by local inhabitants were sometimes crucial to reconstructing the history of various troops. A third advantage of working in this inhabited area was that an excellent map was available. Most of the landmarks I use to describe langur movements already had a name. Wherever possible, the local name, usually Hindi or else a carryover from the British occupation of the hill station, was used. Site names such as Nakhi Lake, Windermere Lake, Eagle's Nest, the Phiroze, and so forth are the names used by local inhabitants (fig. 4.5).

Despite the advantages enjoyed by langurs—and their observers—from living in close association with people, as with most Indian wildlife, the natural habitat of langurs is gradually being destroyed. The vast treeless plains of northern India, which in historical times were covered with equally vast tracts of forest, attest to the inherent conflict between survival of large numbers of people and preservation of forest habitats. Supposedly, the remaining forests

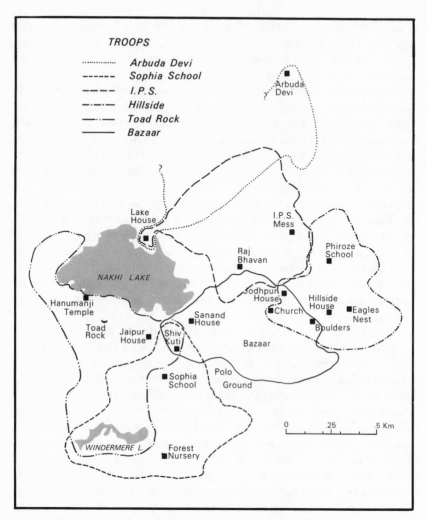

Fig. 4.5. Home ranges of six troops in the vicinity of Abu.

of Mount Abu are protected by law and no live wood may be cut without permission. Enforcing such legislation, however, is nearly impossible when so many tribal people depend for a living on cutting firewood. Nearly every part of the forest bears scars where limbs have been hacked off in the process of "making" dead wood. The highly adaptable—and, of course, sacred—langurs have withstood depradations to the environment more successfully than other species. Nevertheless, should the special status accorded langurs change, it is unlikely that they could survive in the populated areas of India.

Hindu convictions forestall the outright killing of langurs; rumors of a langur poisoning two years before my arrival at Abu were recalled with obvious disapproval. On two occasions, I was beaten with a cane by an elderly holyman who mistook my marking langurs with red ink for intended murder. Nevertheless, stones, slingshots, airguns, and canes are used without compunction (and often with glee) to drive monkeys away from gardens or rooftops. Stoning langurs is a public pastime for young boys and others.

Increasing secularization, combined with the inescapable fact that langurs do cause damage, is undermining the unique protection that monkeys in India enjoy. In addition to crop raiding (Mohnot, 1971a), langurs' wasteful manner of breaking off branches when they feed damages ornamental plants and other property. According to McCann (1928), a traditional form of Indian revenge was to spread grain on an enemy's roof so that langurs would come and pry up the tiles. McCann describes an unmendable feud between two villages that was occasioned by the decision of one town to rid itself of langurs. The town's entire monkey population was loaded onto a freight car. When the train was accidentally derailed beside a neighboring town, the langurs were released and the feud inaugurated. Belief in the tile menace obviously persists today. On a visit to the south of India in 1972, I was told that if I wanted to find langurs, I must ask the tile workers, who "follow the langurs everywhere in order to profit from the damage that they do." In fact, near Dharwar I was told that the tile workers themselves release langurs in order to make more business!

In addition to ruining roofs, langurs invade kitchens, break food containers, and occasionally cause serious accidents. I once saw a langur start a fire by reaching through a window in the bazaar and knocking a tin of kerosene into a brazier. The resentment caused by the antics of monkeys is widespread. A recent article (March 9, 1973) in the *Times of India* was headlined "Monkeys Are a Big Menace to Farming." An excerpt from this article reads, "Sporadic attempts were made in the past to kill monkeys by offering rewards but 'religious sentiments and other reasons prevented their killings [sic],' Mr. H [general secretary of the Mysore farmer's forum] complained." The article added that "many felt the ideal remedy would be to export monkeys for foreign exchange."

Morbidity and Causes of Mortality

The biological fitness of any primate—a potentially long-lived animal capable of producing offspring throughout most of its life—has two components: the reproduction of offspring who in turn survive and reproduce, and the survival of the parent individuals long enough to reproduce again and to contribute to the survival of previous offspring. This study is primarily concerned with the first aspect of fitness—how langurs directly maximize the number of surviving offspring. But the more immediate issue of survival itself must also be considered.

Two hundred and twenty-five distinct langurs were inspected at close range for noticeable pathologies. This sample population included 66 males, 118 females, and 41 immatures whose sex was not always determined. It should be kept in mind that members of this sample were not equally represented in the number of days they were seen. Whereas the same females were seen again and again, many juvenile-to-adult males were observed during only one study period. There would then be a bias towards recording an injury for such females. In spite of this bias, I believe that the sample is adequate to make some important distinctions concerning the likelihood of injury for various age and sex classes.

Pathologies recorded are itemized by age and sex in table 4.3. Arbitrarily, these have been divided into five categories: (1) injuries, classified by type and site; (2) electrical burns; (3) anomalies presumed to be congenital; (4) skin and eye disease; and (5) serious illness. Because of the difficulty of detecting illness under field conditions, this last category is probably not realistically represented by the one case observed, a juvenile male who suffered from temporary paraplegia.[1]

Most injuries were observed after the fact. Almost invariably, though, whenever an injury causing a wound was witnessed, either at Abu or elsewhere (Sugiyama, 1965b; Mohnot, 1971b; Hughes, 1884), the agent of injury was another langur. At Abu where predation is infrequent, the following types of injury can be attributed to a conspecific with some degree of confidence: disfigurement or wounds in the facial region; torn ears or ears that have been bitten off; bite wounds in the scalp; wounds on the body; and cuts on the tail (color fig. 5). Within the sample inspected, sus-

---

1. In 1974, the partially paralyzed juvenile Quilt was taken by me from his troop. Quilt subsequently recovered in the home of a local inhabitant and was returned to his natal troop (Nirmal Kumar Dhadhal, personal communication).

TABLE 4.3. Cumulative catalogue of pathologies observed at Abu through June 1975, listed by age and sex for 225 animals inspected.

Pathology	Adults		Subadults		Juveniles			INFANT-2s		INFANT-1s			Totals
	M	F	M	F	M	F	Uᵃ	M	F	M	F	Uᵃ	
INJURIES													
Face, nose, & mouth region													
Disfigurement	4	0	0	0	0	0	0	0	0	0	0	0	4
Superficial wounds	9	5	0	0	0	0	0	0	0	0	0	0	14
Bloody nose	0	1	0	0	0	1	0	0	0	0	0	0	2
Ears													
Bitten off	0	0	1	0	0	0	0	0	0	0	0	0	1
Major tear	4	3	0	0	0	0	0	0	0	0	0	0	7
Minor tear or pierce	4	1	0	0	0	0	0	0	0	0	0	0	5
Scalp													
Bite wounds	0	0	0	0	0	0	0	0	1	1	0	2	4
Tail													
Broken	3	1	0	0	0	0	0	0	0	0	0	0	4
Truncated	1	2	0	1	1	0	0	0	0	0	0	0	5
Cuts or scarsᵇ	6	2	0	0	0	0	0	0	0	0	0	0	8
Limbs													
Loss of arm or paw (due to electric shock?)	0	1	0	0	0	1	0	0	0	0	0	0	2

(continued)

TABLE 4.3. Cumulative catalogue of pathologies observed at Abu through June 1975, listed by age and sex for 225 animals inspected. (continued)

Pathology	Adults		Subadults		Juveniles			INFANT-2s		INFANT-1s			Totals
	M	F	M	F	M	F	U[a]	M	F	M	F	U[a]	
Hand and foot wounds[b]	6	0	0	0	0	0	0	0	0	0	0	0	6
Probable broken limbs or digits	0	2	0	0	0	1	1	0	1	0	0	0	5
Abrasions	0	1	0	0	0	0	0	0	1	0	0	0	2
Temporary limp	0	2	0	0	0	0	0	0	0	0	0	0	2
Body													
Deep wounds	2	2	0	0	0	0	0	0	0	1	0	2	7
Abrasions	0	1	0	0	0	0	0	0	0	0	1	0	2
Injured pelvis (due to fall?)	0	1	0	0	0	0	0	0	0	0	0	0	1
Probable concussion due to fall after electric shock (died)	0	0	0	0	0	0	0	0	1	0	0	0	1
Total:	39	25	1	1	1	3	1	0	4	2	1	4	82

										Total
CONGENITAL ANOMALIES?										
Deformed paws or toes	0	2	0	0	0	0	0	0	0	2
Mouth area	1	1	0	0	2	0	?	?	?	4
Total:	1	3	0	0	2	0	?	0	0	6
SKIN AND EYE DISEASE										
Ectoparasites	0	2	0	0	0	0	1	0	1	4
Mange	0	16	0	0	—	0	—	0	0	16
Cataracts	0	4	1	0	1	0	0	0	0	7
Chronic "conjunctivitis"	0	1	0	0	0	0	0	0	0	1
"Angioma"	0	1	0	0	0	0	0	0	0	1
Total:	0	24	1	0	2	0	1	0	1	29
SERIOUS ILLNESS[b]										
Temporary paralysis	0	0	0	0	1	0	0	0	0	1[b]

a. Sex unknown.
b. Probably underestimated.

ceptibility to injury varied according to age and sex. While 39 of 48 adult males (81 percent) showed one or more signs of having been injured, only 23 of 70 females (33 percent) had any visible scar or injury on their faces, bodies, or tails. This tally does not include minor scrapes or scratches, but males also exhibited more of these. By contrast, no fully adult male was ever seen with skin or eye disease, while at least 24 adult females suffered from such pathologies, some of them chronically.

At Dharwar as well as at Abu, injuries inflicted by one adult on another were occasionally quite serious. It seems probable that such fights would occasionally result in death. Nevertheless, the only two recorded instances of adults actually being killed by conspecifics come from the nineteenth-century account quoted in chapter 2. Though I personally find the events there plausible, circumspection is called for. There is no dearth of observations, however, concerning fatal injuries inflicted by adults upon infants. According to reports from Dharwar, Jodhpur, and Abu, this is a routine event, and at Abu infanticide constituted the single greatest cause of mortality.

About half of the langurs born at Abu during the five years of this study died during infancy.[2] This figure includes all infants who disappeared from their natal troop before the age of weaning and were presumed dead. The likelihood of survival of any given infant, however, varied radically with troop stability. Infant mortality ranged from a low near zero for the Toad Rock troop during the stable years between June 1971 and early 1973, to a high of 83 percent mortality (10 of 12 infants born) in the very vulnerable Hillside troop between June 1971 and February 1973, a period when the troop changed alpha males at least three times and was invaded at various times by no fewer than seven different males. The somewhat lower mortality in the Bazaar troop during the same period (about 50 percent, or 10 of 19 infants), almost surely reflects the ability of the usurping male (Shifty Leftless) to restore order soon after his 1972 take-over—that is, to keep other males out of the troop. The attrition rate of infants born in the Hillside troop was so great that by 1974 I began to wonder if this troop composed of six females growing older every year would not eventually disappear. When I returned in the spring of 1975 the oldest female (Sol) had disappear-

2. Obviously, any discussion of mortality must be limited to infants whose existence I knew of. If a number of infants were born and died during my absence from Abu, I would be grossly underestimating mortality.

ed, but two of the five remaining females were accompanied by daughters over seven months of age. These two infants were nearly weaned, and their prospects for survival seemed good. Three new infants (two females and a male) were born into Hillside troop that spring. By October 1975, when Harvard biologist James Malcolm visited Abu, one of the older female infants, and a second younger infant (Harrieta's daughter, born April 28) had disappeared. A new male, recognizable by a two inch gash in the right side of his throat, was traveling with the troop. By taking into account births in the last year of my study, the prognosis for Hillside troop seemed slightly improved. Of seventeen infants that I knew about, six survived.

For langurs living within the vicinity of the town of Abu, electrocution is the second most serious hazard. Abu is crisscrossed by live electric wires that, during the rainy season, are draped with the bloated bodies of fruit bats who inadvertently fly into them. Much more knowledgeable about these wires than other treetop denizens, langurs will gingerly test wires with their fingers before catching hold. Nevertheless, in the last six years, townspeople have witnessed the electrocution of three langurs. Two of these (one adult male, one female) died before this study began; the third was the fourteen-month-old daughter of a Bazaar troop female, Wolf. In addition, two females in the study population were missing forelimbs, probably as a result of electric shock. One, a juvenile female in Bazaar troop (Guaca) was observed in August of 1972 just days after her injury; few other hazards could explain her charred and atrophied limb. Gauca disappeared temporarily from her troop. When she rejoined it later in the month, she had lost her arm above the elbow. Though Gauca remained small for her age, within seventeen months of this accident she exhibited estrous behavior and copulated with the alpha male. In January of 1974 she gave birth to a daughter. Although Gauca was an adequate mother, it may not have been purely coincidence that when sighted fifteen months later, the daughter of this handicapped mother had herself had a crippling accident that left her right arm twisted above the wrist. The second animal missing a limb was an older multiparous female (Pawless) who lost her arm below the elbow. Pawless had little difficulty caring for either of her two daughters born during the study. Despite their handicaps, both females handled themselves well, depending to a great extent on ricocheting with their hind quarters off trees and buildings, and both participated vigorously in conflicts with aliens.

In six cases, I suspected that observed abnormalities were congenital defects. These included one adult female (Pawlet) who had a "club" right hand with one digit emerging; an adult female (Oedipa) who was missing toes on her left rear foot; and four animals in the same troop (Harelip, the alpha male; a young adult female; and probably two juvenile males) with a defect such that the tongue was chronically protruding (fig. 4.6). Neither of Pawlet's two daughters nor any of Oedipa's three offspring shared the maternal defect. In the case of the protruding tongue, Harelip was suspected of fathering other individuals with the trait. If Harelip's abnormality is in fact genetic, it may provide a marker for his line.

Fig. 4.6. Even at a distance, Harelip could be recognized by the pink tongue bulging out from between his lips.

Identifying Individuals

Vital to any behavioral study is the ability to recognize individuals and to keep track of them over time. It would hardly have been diplomatic for a foreign tourist to capture and permanently mark Abu's revered langurs. Instead, detailed written and sketched records were kept for each animal. By 1975 some 242 langurs—183 members of bisexual troops and 59 extratroop males—had been encountered. Of these, 78 were individually known langurs. These known individuals could be divided into two classes: 36 marker individuals with some permanently distinctive feature and 42 acquaintances which could be recognized within a given social or geographic context. All marker individuals could be recognized from study period to study period on the basis of one or more diagnostic traits—the pattern of a torn ear; a distinctive ear shape; a severed, broken, or deeply scarred tail; a chronic eye infection or "cataract"; a missing or broken limb or digit; a very distinctive mole; facial scars; and in two cases, a very unusual facial expression. Most acquaintances could also be identified at a later date, provided that they remained in the same area and in the company of the same individuals, usually including one or several marker individuals. Useful indices for identifying acquaintances included age; rough assessments of parity; associates (especially in the case of infants and their mothers); and subtle physical differences, especially ear shape, coloration, moles, facial expressions, shape of tail tip, and in the case of females, peculiar depigmented blotches between their haunches. Within the same study period, temporary identification of animals was facilitated by squirting them with colored stains. These spots of color were important for identification of langurs partially obscured by foliage whose identifying traits could not readily be seen. During the dry season, mange is relatively common and depilated patches were also used for quick identification of individuals. Despite the number of slight physical differences between individuals, it was sometimes difficult to differentiate young, same-sexed animals of approximately the same age once they were weaned. Wherever such confusion occurred, it is indicated either in the text or in a note to the relevant table.

Almost all the marker individuals and acquaintances were at one time members of one of six troops that ranged in the vicinity of Abu town. Behavioral observations were concentrated on three of these: the Bazaar, Toad Rock, and Hillside troops. By 1973 all adults, sub-adults, and dependent immatures (but not all juveniles) in these troops were individually known. The compositions of Bazaar and

Toad Rock troops between 1971 and 1975 are provided in tables 3.6 and 3.7; changes in the composition of Hillside troop are given in table 4.4. Other troops about the town and the Chippaberi bus stop troop were less well-known. Those about the town included the Arbuda Devi Temple troop (containing 19 individuals in 1971), the I.P.S. troop (25 in 1971), and the School troop (table 4.5). Only some animals in these five troops were individually identifiable. Of the 59 extratroop males recognized, 34 of them made one or several shifts between membership in bisexual troops and membership in all-male bands during the course of this study. Because of the constantly fluctuating membership of male bands, it was rarely possible to reidentify these animals on the basis of social context. There were, however, eight marker males who moved between bisexual troops and male bands and to some extent it was possible to follow their careers from year to year.

Aging a Langur

Because age figures so prominently in my interpretation of langur behavior, it is important to note that an estimate of each animal's age was made as soon as possible after initial identification. Hence, an individual's age was usually assessed prior to, and therefore independently of, most behavioral observations about him or her.

Changes in skin and coat color make it relatively easy to determine

TABLE 4.4. Changes in the composition of Hillside troop between July 1971 and June 1975.[a]

Date	Adult male	Adult female	Infant	Juvenile
July 1971 (total = 15)	Mug	Bilgay Itch Oedipa Harrieta No-name Pawless Sol	1. 2. 3. 4. } twins 5. } 6.	Sancho (m)
August 1971 (total = 8)	Shifty	Bilgay Itch Oedipa Harrieta Pawless Sol		Sancho (occasionally present)

TABLE 4.4. Changes in the composition of Hillside troop between July 1971 and June 1975.[a] (continued)

Date	Adult male	Adult female	Infant	Juvenile
June 1972 (total = 11)	Shifty alternates with Mug	Bilgay . . . . Mira	Mira (f; 3 months)	
		Itch . . . . Scratch	Scratch (m; 3-4 months)	
		Oedipa . . . . Virginia	Virginia (f; 7-8 months)	
		Harrieta . . . . . . Harry	Harry (m; 6-7 months)	
		Pawless		
		Sol		
February 1973 (total = 14)	Shifty alternates with Mug Righty Ear Kali No-No Man Bluebeard Pequeno	Bilgay (pregnant) Itch Oedipa Harrieta Pawless . . . . Pawla Sol	Pawla (f; 3-4 months)	Harry
December 1973 (total = 9)	Mug	Bilgay . . . . Miro (m; ca. Itch 10 months) Oedipa Harrieta Pawless Sol	Miro (m; ca. 10 months)	Harry
April 1975 (total = 10)	Righty Ear	Bilgay (pregnant) Itch . . . . 1. Female (ca. 12 months) Oedipa . . . . 2. Female (ca. 7 months). Harrieta . . . . . . 3. Female (born Apr. 28) Pawless . . . . 4. Female (2 months)		

a. M indicates male; f indicates female. Dots connect females with infants presumed to be their offspring.

TABLE 4.5. Estimated changes in the composition of School troop between July 1971 and January 1975.

Date	Adult male	Adult female	Infant-1	Infant-2	Juvenile	Subadult
July 1971 (total = 19)	Harelip	8	0	8	0	2 females
June 1972 (total = 16)	Harelip	7	0	2	6	0
February 1973 (total = 16)	Harelip	7	4	0	3	1 male
January 1974 (total ≈ 21)	Harelip	7	1	2?	6?	2 males 2 females
April 1975 (total ≈ 17)	Harelip	7	2	6		1

the age of langurs in infancy. At birth, langurs are covered with feathery black-brown lanugo. The pale skin is nearly transparent, and the blood visible beneath the skin tints the face, ears, and the bottoms of the infant's feet flamingo-pink. Pigmentary deposit in the skin begins about a week after birth; the skin on the wrinkled little face turns ashen and then brown. By the third month the face is completely black. Coat color begins to change slightly later than the skin. Between the third and fourth month, the hair on top of the head turns white, beginning at a spot in the center and gradually expanding; at the same time the infant displays a little white goatee. This gradual transformation from black to cream coat color continues until the fifth or sixth month after birth, when the infant is completely white except for the dark face. A cream-colored infant is referred to as an "infant-2." An infant before this color change is called a "dark infant" or "infant-1." At the time of weaning, langurs are reclassified as "juveniles." The onset of weaning and its conclusion are both variable. Furthermore, an older infant may return briefly and join its newborn sibling at its mother's breast.

Langurs are classed as juveniles from the time that they are weaned until sexual maturity, when they become "subadults." In males, subadulthood is a transition period, marked by gradual enlargement of the descended testicles. Given the opportunity, subadult males appear capable of copulatory behavior, though they may have difficulty assuming the correct mounting posture (see p. 299). At this age, the bulging snout of a subadult male—looking as though he had chewing gum stuffed behind his upper lip—can already be distinguished in profile from that of an adult female.

Females are classified as subadults with the onset of estrous cycling. In general these young females are smaller than full-grown adults, weighing from five to ten pounds less than a full-grown adult female. Females do not reach their full body size or exhibit fully adult dentition until after the birth of their first or even second infant. As soon as her first pregnancy can be detected, usually in the last two months before she gives birth, a young female is classified as a full adult (at about four and one-half years). In the case of males, the transition to adulthood is more ambiguous and subjective; there is no clear demarcation between an "older subadult" and a young adult male.

Parity is undoubtedly the most useful tool in estimating the age of a young female. Whereas the nipples of females who have never given birth (nulliparas) are scarcely noticeable dots, those of mothers are prominent. A female who gave birth to her first infant in 1972 would have to be around eight years old at the end of the study (in 1975). Nipple length does increase to some extent with the number of offspring produced; primiparas (who are about to have their first infant or who have had only one infant) are usually distinguishable from multiparous females (who have borne two or more offspring). Within the class of multiparous females, however, it is not yet possible to make fine-grained distinctions. After the first infant is born, pendulosity is no longer a good criterion for aging. Both breast size and the distention of nipples may decrease in a female who has not given birth for several years. The nipples of some very old females most closely resemble those of primiparas.

The life-stage of both male and female adults can be estimated on the basis of facial characteristics and body development. In young animals, the skin of the snout is stretched smoothly and tightly across the jaw structure. The coat color is silver-white and the chest may be almost pure white. During this period females reach their full body weight, about 11 kilograms or 25 pounds (from a sample of 11 multiparas; see appendix 2). Females continue to grow after the birth of the first infant; multiparous females appear to have wider backs. By middle age (beginning at perhaps twelve years) the eyes have become more sunken into the face and are surrounded by creases. By middle age males, too, have reached their full body size, 18 kilograms or 40 pounds (based on a sample of 3 alpha males) and are noticeably stockier than younger animals. At this age the pink pads above the male's ischial callosities—which were mere strips in subadulthood—are pale pink, full and puffy, and, at least in some alpha males, appear to increase in size with age. Females have a corresponding strip of pink skin, but it is narrower than that of an adult male and is marked in the center by the clitoris, a visible pink knob.

In old age, traits apparent in middle age become more exaggerated. Eyes are more deeply ringed by creases (see Napier and Napier, 1967, plate 94 of a *Presbytis pileatus* female at the San Diego Zoo, known to be twenty years of age; see also Hill, 1975). The skin of the snout becomes slightly pitted and more "wrinkled." The nostrils of some old females appear wider and the snout housing the worn-down teeth flatter; the lips may appear to tuck in as they do in the case of toothless humans (see photo of *P. cristata* female, known to be twenty-nine years old; Anon. 1974). Some of the old females at Abu appeared to be much older than any of the males observed; they were bony and generally debilitated. It is possible that their bushier facial hair and the hoary hair on their black hands and feet may have been due to old age. The appearance of three very old females (twenty or more years?)[3] was markedly different from other animals, but very similar to each other: they had flattened, almost snub-nosed faces, wide (gorilla-like) nostrils, and bushy manes of silver hair about their faces (fig. 4.7). These old females' behaviors also resembled one another's but differed from the behaviors of other females in the troop. In both males and females, injuries and other identifying marks tended to accumulate with age. All four females with torn ears, for example, were fairly old.

Cast of Characters

I am no true naturalist. Langur locomotion is awe-inspiring, their shrewd adaptations to local environments remarkable, but if it were not for the fact that langurs interact as individuals, I could never have sustained my interest in them for five years. It was the high drama of their lives, the next episode of the colobine soap opera that got me out of bed in the morning and kept me out under the Indian sun, tramping about their haunts for eleven hours at a stretch. During the five seasons of this study, political histories of the Bazaar, Toad Rock, and Hillside troops were marked by dramatic

3. The colobine longevity record is held by a 30-year-old Colobus monkey at the San Diego Zoo (Hill, 1975).

Fig. 4.7. Sol, old matriarch of the Hillside troop.

power struggles between males, accompanied by more continuous but less obvious fluctuations in the relations between females and by the alternating antagonism and cooperation between these animals. The capsules provided here introduce some of the protagonists in this chronicle and indicate how they were recognized. Catalogues of individuals in each troop appear in tables 3.6, 3.7, 4.4, and 4.5.

In tables 5.4 and 6.5 and figures 6.11 and 6.12, each female is listed by rank along with her reproductive status and weight where these are known.

Males   In most primate species, males are the more conspicuous sex. Troops contain proportionally fewer male adults; males are larger, more likely to bear identifying scars, and their behavior is more boisterous. At Abu, males were frequently the first member of a troop to be individually recognized. Without doubt, the most unmistakable of these was the highly successful middle-aged-to-older patriarch named *Harelip.[4] This large langur (weighing 18.6 kilograms, or 41 pounds) remained the alpha male of School troop throughout the five years of this study. Because his upper lip fails to mesh with the lower, Harelip's tongue can routinely be seen bulging out between his lips (see fig. 4.6).

One of Harelip's frequent preoccupations was his next-door neighbor, *Splitear, alpha male of the Toad Rock troop. These two troops met on an average of once every thirty daylight hours, and whenever they did, the two males displayed flamboyantly and chased one another about; only rarely did these two familiar opponents ever resort to actual fighting. Harelip's "dear enemy" was a stocky, middle-aged male whose name refers to a slice in the top of his left ear. By 1973, Splitear had a second identifying trait, a deep healed wound that creased his face from the center of his forehead down to his upper lip. Splitear remained the alpha male of Toad Rock troop from our initial contact with this troop in 1971 until around the monsoon season of 1973, when a large (17.7 kilograms, or 39 pounds) and lanky young adult named Toad usurped the troop. After Toad's take-over, Splitear and eight young Toad Rock males (believed to be his sons) traveled as a male band within the former Toad Rock troop's home range. By 1975, however, neither Splitear nor any recognizable member of his contingent could be found in the vicinity of Toad Rock troop.

Two of the subadult males who left along with Splitear in 1973 were *Cast-eye, a male 51 months old in January of 1974 who was identifiable by an opaque film covering his right eye and a gap between his front teeth, and *Blind-ear, a youngster who even prior to adulthood had two strikes against him: a sunken and sightless right eye, and a missing left ear that was ripped off on December 28, 1973, probably by another male.

---

4. Starred animals are permanently recognizable marker individuals.

In 1972, a new male named *Shifty Leftless* came into repeated contact with Splitear and Harelip when he usurped the Bazaar troop, whose range abutted theirs. Here Shifty remained until 1975 when he disappeared. Shifty was a squat, muscular, decidedly older male who stared out at the world over a slightly flattened muzzle through piercing eyes set deeply into his face. A "bite" missing from his left ear made him unmistakable in profile. From the rear a full inch or more of puffy seashell pink pads over his ischial callosities could be seen.

Actually, Shifty began his career (as far as our records are concerned) some distance away, in August 1971, when he usurped the Hillside troop. When Shifty first encountered this small troop of seven adult females, a stocky, middle-aged or possibly younger male named *Mug* was in command. Mug's distinguishing features included a tiny slit in the bottom of his left ear, a snout that was slightly askew to the right, and a faint scar on the underside of his tail, half way down. In 1972 Mug was able to return to the Hillside troop, after Shifty had departed for Bazaar troop. In 1973 Mug was joined in Hillside troop by five additional adult males: *Righty Ear*, a young adult with a chunk missing out of his right ear; *Kali sana*, a young male extraordinarily aggressive towards humans; *No-No Man*, a somewhat older male than either Righty Ear or Kali sana; and *Bluebeard* and *Pequeno*, both quite young. Of these five males, only one, Righty Ear, was definitely identified in later years. In 1973-74 Righty was one of three males roaming in the vicinity of Hillside troop. By 1975 a fully adult Righty, weighing 18.7 kilograms (or just over 41 pounds), had taken Mug's place as the alpha male of Hillside troop. Mug had moved on to Bazaar troop. Shortly thereafter, Righty left Hillside for Bazaar troop. Mug disappeared.

The last of the well-known alpha males of the troops about Abu town was *LeGrand*, leader of I.P.S. troop from 1971 to 1974. This smallish, muscular middle-aged or older male could be recognized by his short tail and a tiny pierce through the center of his right ear. By 1975, this exceptionally large troop had split in two, one half led by a very bushy-haired old langur male, and the other half led by a young adult who greatly resembled Toad. LeGrand was never seen again.

Fourteen kilometers below the town, at the Chippaberi bus stop, events very similar to those at the top of the mountain occurred, but the two casts of characters were quite different and almost certainly never met. Clearly recognizable actors in the Chippaberi troop drama were *Misshapen*, an older male with a broken tail, a face full of scars, and a torn right ear; and *Jag, whose only scar—until the events of August 1971—was a V-shaped scar half-way down his tail on the dorsal side.

Females   Identification of the more difficult sex depends heavily on context. Fortunately, it is exceedingly rare for langur females to switch troops; they are listed here by the troop that they remained with throughout this study.

Toad Rock Troop. *Pawlet epitomizes the intratroop vicissitudes of female life and more than any other female at Abu raises questions about how and why changes in female rank occur. This middle-aged female, readily identifiable by a deformed right paw (a fist fused into a club with one finger emerging), was able to displace every other female in the Toad Rock troop in the monsoon season of 1972. Pawlet could also displace most immature males in the troop and engaged in fierce spats with the alpha male, Splitear, though he remained dominant to her. By 1973-74, however, Pawlet (who was at that point accompanied by a weaning infant) had plunged in rank so that she interacted very little with other females and never displaced them. As of 1975, Pawlet rose again to the upper half of the female hierarchy.

The female who supplanted Pawlet in 1973 was *Mopsa, another multiparous, middle-aged female with a distinguishing tear at the top of her left ear. By the end of 1973, Mopsa was herself replaced as the alpha female by Mole, a low-ranking subadult female at the outset of the study who gradually rose to the position of beta female in early 1973 and to alpha position by the end of that year. Mole has a scarcely visible slit in the bottom of her right ear, but in 1973 and 1974 was unmistakable because of her sleek silver coat and the smooth black skin stretched tightly across her muzzle. By 1975 Mole had fallen in rank below Pawlet, a young adult female, and four young subadults. Mopsa and her infant son had disappeared from the troop, and I presumed them dead.

Following precisely in Mole's pattern, a low-ranking immature from 1973-74 rose to the alpha position in 1975. This was Hauncha. Two other subadults, Pandy and Handy, followed her lead, and ranked just below her in 1975.

Sandwiched between these changes were two consistently middle-ranking, middle-aged females named *T.T. (for Tied Tail), and *I.E. (for Infected Eye). T.T. could be identified by a tail that was truncated half way down, I.E. by a chronically infected right eye, a hole in her left ear, and the occasional appearance of a growth or tumor at the bottom of her stomach.

Near the bottom of the Toad Road hierarchy was Circle Eyes or *C.E., an old (but still reproducing) female with deep semicircular creases beneath her two bluish-tinged, bleary eyes. Though C.E. was very aggressive towards me, threatening and slapping at me with little provocation, she was skittish about her troopmates and avoided interactions with them, foraging out of sight of any other langur for up to fifty minutes at a stretch.

Bazaar Troop. Elfin, a young adult with peculiarly peaked "elfin" ears, rose to the top of the female hierarchy about the time she first gave birth and remained there throughout the study. As in Toad Rock troop, several subadults were gradually rising in her wake; these included *Breva, who lost part of her tail as a juvenile in August 1971, Junebug, and *Guaca, who lost her right arm above the elbow in August of 1972.

Two females in Bazaar troop were blind in their right eyes: *Overcast, a middle-aged multipara, and *Wolf, a middle-aged-to-older female who routinely lost much of her facial hair each dry season. The two oldest females in Bazaar troop were *Short, a bushy-haired matron with a short tail and a penchant for foraging alone, and *Quebrado, a female with a kink in her tail.

Hillside Troop. The oldest female of any seen at Abu was almost certainly *Sol, whose name derived from her seemingly solitary disposition. Sol's wide nostrils and flattened snout gave her a gorilla-like appearance (fig. 4.7). By 1974 she had accumulated a barrage of small scars on her long tail, and almost never traveled with the troop during the day, though she could sometimes be seen on a high rocky outcrop not far from the troop staring into the distance like a lonely sentinel. After that season, Sol was never seen again. The other striking individual in the Hillside troop was *Pawless, a middle-aged-to-older female who had lost her left forearm but who was nevertheless a competent participant in troop affairs.

Pros and Cons of a Social Existence

Early in the morning the troup goes out to feed, the leader keeping a vigilant eye on all stragglers though he does not appear to trouble much about the yearlings, but in a case of danger comes to their rescue. To illustrate this point I shall narrate a little incident which occurred while out collecting. I shot a female which was carrying her baby (probably a couple of months old). The body of the mother fell to the ground but the young remained on the tree. The report of the gun dispersed the others. The baby was most active but my attendant and I managed to isolate it on a tree, as the span of the next tree was too great. I sent my man up the tree to try to catch it. When it saw him it began to squeal. The cries of the little one soon brought the others around once more. I fired a couple of rounds to disperse them. Eventually the leader of the troop came forward and, in spite of every attempt to drive him away, he jumped across to the tree, but before he could get to the young one I brought him down. We finally caught the baby. This shows the strong social attachment and protective instinct of the overlord of the troup.

Charles McCann, 1933

Altruism, or the unselfish concern for the well-being of others, is a uniquely human edification. Its analogue in the animal world takes two forms: self-sacrificing acts that benefit close kin, and risk-taking rescue or generosity that carries a high probability of future reciprocation (Hamilton, 1971; Trivers, 1971). A langur troop leader who rescues an infant benefits the endangered animal; he also preserves an individual who very likely replicates a minimum of one-half of his genes. Such daredevilry as McCann reports is an extreme case. Most solicitude towards troopmates exacts a lower cost. Gains to the recipient are cumulative and inconspicuous: the daily rewards of social life that offset the omnipresent disadvantages of living among conspecifics who compete for the same resources.

The most routine advantage of group living are alarm cries by whichever animal senses danger first. A staccato bark from an adult, a fear squeal from an immature, jolts the troop into a state of readiness, and, if necessary, flight. False alarms, especially by immatures are common, but langurs seem to have little difficulty distinguishing real danger, since alarm squeals from youngsters are often ignored. Usually when langurs are truly threatened, they are on the ground. Hence giving an alarm signal does not make the sender any more conspicuous than he or she already is. At little extra cost to self, the sentinel has created general confusion as troop members scramble for safety, contributed to the survival of relatives, and participated in a system in which reciprocity may be an essential element. If groups of langurs are more noticeable, they are also more wary. Like many relatively defenseless animals, langur security relies on the vigilance of many noses, eyes, and ears.

Since predation upon langurs has only rarely been described, it is not always clear exactly what langurs are on the lookout for. The rarity of observed cases may be due to the deterring presence of human witnesses, and almost certainly belies the actual incidence of langurs being eaten. When George Schaller studied feline predation in the game parks of northern India (1967), he found that 6.2 percent of 335 samples of tiger feces contained langur hairs, as did twenty-seven percent of 22 leopard samples. Langur remains have also been found in the scats of the Indian *dhole* or wild dog (J. Malcolm, personal communication). In the Gir Forest, the Indian primatologist Rahaman watched as a hawk (*Accipter sp.*) swooped down upon an infant langur, but the attack was thwarted by an adult male (1973). S. M. Mohnot observed a kite (*Milvus migrans*) fly off with the corpse of a dead infant (1971b).

Within the populated environs of Mount Abu, the only reported predation in recent years has been by *Panthera pardus*. Many people claim to have seen encounters between leopards and langurs. The most reliable account was told to me by Fethe Singh, game warden at Abu, in 1971. Around 11:00 one night a leopard was stalking beneath a tree full of alarmed langurs. One langur decended from the tree, was caught, and eaten. This odd episode echoes accounts of fatal descents in the presence of predators related by early British naturalists. Why langurs do this remains a mystery.

In the course of my study, forty-four animals disappeared from four troops and were never seen again. Only nine of these disappearances could not be explained by political changes within the troop that routinely coincided with the expulsion of males and the killing of infants. In only one case did an adult female disappear at the time of a take-over. Of the nine unexplained absences, eight were adult females, three of whom were quite old. The ninth was the infant son of one of these females. The missing females could have been victims of predation, disease, old age, or some combination of these. Another possible victim of predation was the middle-aged Toad Rock female Pawlet who lost the end of her tail during the night of January 23, 1974, while her troop was roosting in trees in a relatively unpeopled portion of its home range. When I encountered the troop at 7:20 the next morning, the alpha male was uttering continuous grunts and the whole troop was unusually active. By 7:45 the langurs had moved some distance from their sleeping site— an exceptionally early start for langurs on a cold winter morning. It is possible that during the night an unsuccessful predator made off with the tail end of an elusive meal.

Langurs around the town are also subject to harassment by dogs and men. Though typically nonlethal, such harassment forces langurs to spend much of their time in or near trees, on walls, and on rooftops. When undisturbed, langurs may spend long hours feeding and grooming on the ground. So agile are they at bounding from the ground back into the trees that, if forewarned, the long-legged langurs easily evade danger. So confident are langurs of their ability to escape that at Abu an adult male may descend from the safety of the trees to slap teasingly at a dog before catapulting out of reach again. On several occasions over a two-day period, I provoked competition over bits of bread between a rather tyrannical troop leader (Splitear) and an unusually sleek and powerful local dog. In the first hours, Splitear's feints kept the dog at bay, but gradually the dog grew bolder. By the second day, Splitear watched from above as the dog obtained all the food rewards. Elsewhere, dogs may pose a more serious threat. A subadult male langur at the Gir Forest was caught and eaten by two village dogs while the adult male and other troopmates watched from the trees above (Rahaman, 1973). Packs of village dogs were responsible for the death of a number of langurs at Ripley's study site at Polonnaruwa (personal communication).

Descriptions of langurs in the current primate literature depict adult males who are "aloof" and lacking in fatherly concern (for example, Mitchell and Brandt, 1972; Martin and Voorhies, 1975). Aloof these males may be, but they nevertheless take risks in defense of infants born in their troop, as described in McCann's 1933 account. A similar story was told to me by townspeople in the bazaar at Abu just hours after the incident occurred. An infant had been shocked by an electric wire in the bazaar and had fallen to the ground without the mother (Wolf) seeing what had happened. For more than half an hour, Shifty Leftless, the single adult male in the troop, stood guard over the corpse, refusing to allow any human to approach until the mother herself had appeared on the scene and retrieved it. On the day following the accident, I attempted unsuccessfully to examine the body, which Wolf was still carrying with her but had temporarily laid aside. Attacks from the surrounding langurs forced me to drop the body. On the second day following the infant's death, when the corpse had been altogether abandoned by its mother, Shifty Leftless unexpectedly charged me after two females gave alarm cries unrelated to anything I was doing. Hurling two notebooks and a pen at the charging male, I ignominiously enhanced his bluff(?) by scurrying away with Shifty in pursuit. Apparently my behavior on the day before had caused the langur to classify me as a predator. Sugiyama was similarly attacked by a troop leader each time he approached langur troops at Dharwar holding a captive infant. The male attacked regardless of whether the infant had been captured from his troop or from another (1965a), possibly because the male could not differentiate this at a distance. In the close call between a Gir Forest infant and a hawk mentioned above, the mother had left her infant on the ground while she herself sought safety in the trees. When the hawk swooped down, the adult male leapt to a low-hanging branch just above the infant, and in doing so caused the branch to sway, grazing the hawk's back and startling him away.

Females as well as males take risks in defense of troopmates. By itself, however, the likelihood of close genetic relatedness does not predict when intervention will occur. In some cases, allomothers, or females other than the mother, play the most daring roles in the defense of a threatened infant; other times, they desert it. Even the infant's own mother may desert it if the cost of intervention is high or if the infant is unlikely to survive. Mothers confronted with an undeterrably infanticidal male may cease to defend their infant or abandon a wounded infant altogether. Presumably, a mother with the potential to produce several more offspring in the course of her lifetime minimizes her loss by terminating investment in the luckless infant and trying again. A female with an infant of her own is even more likely to desert some other female's baby if danger should threaten while both infants are in her custody. On at least two occasions at Abu, mothers rescued their own infants from human harassers (attempting in one case to stone, in the other to capture, the langurs), while leaving behind a second baby that was in her vicinity.

Similar episodes may have prompted Frank Poirier to write of Nilgiri langurs that "other troop members shared the mother's apparent lack of concern for her infant's safety. Rather than aid an infant caught in a precarious position, a nearby animal often left the scene, especially if the infant began to vocalize in distress" (1968b). But the unmitigated selfishness described here does not always tally with the heroic record of old females among hanuman langurs.

Defense of other troopmates by females was seen primarily in four contexts, involving dogs, human harassers, langurs belonging to other troops, and infanticidal males. On seven occasions a female descended from a position of relative safety in the trees to threaten, slap at, or in one case (involving Pawless) to momentarily grapple with a dog who was harassing the troop. The defending female was

Pawless on three occasions, Sol on one, Quebrado on two, and Pawless, Harrieta, and Sol acting in concert on one. These defenders all tended to be old animals. Only once was a young female, Bilgay, seen to take on a dog, and only after it had forced her from a pile of provisioned chickpeas; she was apparently defending her own food source. Old females also appeared to be least hesitant to charge harassing humans. Pawless charged a watchman attempting to chase Hillside troop out of a garden and chased a small boy who had hit her with a stone as the troop crossed a road. Old Short from Bazaar troop and an aged School troop female named Chip, who suffered from a large angioma on her flank, would also, on little provocation, chase children who were teasing any members of their troops. Short, together with Quebrado, also played a key role in chasing away a female from another troop whose infant had been kidnapped by a Bazaar troop subadult. Interestingly, though Short and Quebrado were foremost in preventing the mother from retrieving her stolen infant, neither female subsequently exhibited much interest in handling the stolen baby. Other females, in particular those who had never had an infant of their own, appeared to benefit most from holding and inspecting the stolen infant. Similar selflessness on the part of old females could be seen when two troops met at their common border and engaged in an aggressive encounter. Both young and old females participated, but the rewards of these encounters, whenever specific items were at stake, went to young females. For example, when I aggravated an encounter between I.P.S. and Bazaar troops by putting out food between them, it was old Short who advanced against I.P.S. troop and drove them back. But despite the fact that Short had been almost single-handedly responsible for capturing the prize (a pile of chickpeas), the food was immediately monopolized by Elfin, the Bazaar troop's young alpha female. Short obtained none.

The most spectacular instances of intervention by old females involved infanticidal males. On numerous occasions when Mug attacked Hillside troop infants, females other than the mother intervened, threatening, chasing, and even grappling with the male. Sol (on twelve occasions) and Pawless (on seven) were the main participants in this defense. In 1975, when Righty Ear made repeated attacks on Bazaar troop infants, Quebrado was their staunchest defender.

Apart from such dramatic rescues, old females exhibited little concern for infants and interacted very little with troop members generally. The females with the greatest interest in infants, who literally could not keep their hands off them, were young animals who had never given birth themselves, or who were about to. Though inexperienced in infant care, these nulliparas were so eager to take and retain infants, and to prevent retrieval by the mother, that in general they handled their charges with great care. This babysitting arrangement seemed to give the young females an opportunity to try out their maternal skills on another female's baby; it also gave the mothers freedom to forage unhampered by the responsibility of her infant's safety or the cumbersomeness of a newborn. Assistance in caretaking may be especially important in the case of twins, which are not infrequent in the langur population at Abu. Three pairs of newborns were present in two troops (I.P.S. and Hillside troops) during the course of the study. Additional benefits to the mother-infant pair include the possibility that an allomother would adopt an ophaned infant. Adoption has been reported now for several primate species, including rhesus and Japanese macaques, chimpanzees, and in one case, hanuman langurs (Boggess, 1976). Almost invariably adoption has been by a close relative, for example, a sibling (Sade, 1965; Lindburg, 1971) Van Lawick-Goodall, 1968) or grandmother (Itani, 1959), or else by a female primed to be a foster mother by intensive and intimate prior contact in a cage (Rowell, Hinde, and Spencer-Booth, 1964). Quite possibly, intimate contact in captivity simulates the conditions of family life in the wild. Bonnet macaques, for example, are extraordinarily amenable to adopting orphans in captivity (Rosenblum, 1968) and are also characterized by closeknit, presumably family, clusters in the wild, where it would be rare for a female to have an opportunity to adopt an unrelated infant.

The fate of an orphan langur has never been described, although there is little question that an abandoned infant would be picked up and carried by one or more females. Within the troops studied at Abu, the average time that an infant abandoned by a caretaker was left unattended was under two minutes. The real question, however, is what sort of treatment an infant without a mother would receive, and especially whether or not an unweaned orphan would be allowed to suckle by a lactating foster mother. Unweaned infants kidnapped from other troops sometimes starve to death, and orphans might face a similar end. The reason is that hanuman langurs do not ordinarily suckle another female's offspring. I witnessed one exception in which a mother briefly suckled an infant kidnapped from another troop in addition to her own infant, but moments later the allomother rebuffed the kidnapped infant. A similar episode in which a mother hanuman langur allowed a second infant to nurse from her at the same time as her own was reported at the San Diego Zoo. The allomother suckled the second infant for two and one-half minutes before it was retrieved by its own mother (J. McKenna, personal communication 1975). At Abu older infants attempting to pirate milk from allomothers were chastised, sometimes fiercely. Typically, a slightly older infant will attempt to join a dark infant at its mother's breast, but the intruder is met with threats and cuffs from the mother. A few yearlings in the process of being weaned may make a career of milk-stealing attempts, passing from one female to another, rejected by each in turn. Though females do not normally allow another female's offspring to nurse, her own older offspring may return and suckle from time to time after the birth of a new sibling.

Even if a female's own offspring has died or is not nursing at the time, she may refuse to allow a substitute to suckle. For example, on the day after the infant Scratch was seriously wounded by an adult male, Scratch ceased feeding. Other Hillside troop infants, however, tried unsuccessfully to avail themselves of his mother's milk. The day after the accidental death of Wolf's son, she repeatedly treatened away juvenile and older infant milk pirates even though Wolf appeared to suffer considerable physical discomfort from not nursing; throughout the day, she pulled at her own nipples. This reluctance to share milk with troopmates is not altogether surprising; suckling is an expensive form of solicitude because it delays a female's next conception.[5] The fact that Wolf refused to continue suckling after the death of her own infant probably reduced the time elapsing before she gave birth to another. In contrast, at Boggess' Himalayan study site, an infant whose mother was still alive in the troop was adopted by another female whose own infant had died. The foster mother had begun taking and holding the borrowed infant during pregnancy, before her own ill-fated infant was actually

5. The effect of an infant's death on the mother's subsequent birth interval has been described for rhesus and Japanese macaques (Hartman, 1932; Tanaka et al., 1970). To date the best documentation linking a delay in subsequent conception to lactation comes from human studies (Douglas, 1946; Smith, 1960; Bonte and van Balen, 1969; and Ring and Scragg, 1973). One of the clearest examples of the correlation between maternal investment and subsequent birth intervals is provided by the Ring and Scragg study of human fertility in rural New Guinea. It shows that after stillbirths or infant deaths, mean birth intervals increased with the age of infant at death: 1.31 years after a stillbirth; between 1.45 and 1.74 years if the infant was under six months; 2.48 years if the infant was between one and two years; and 2.64 years if the infant survived more than two years.

born (J. Boggess, personal communication). The explanation for this curious episode must await further information (Boggess, in preparation).

The self-interested limits to solicitude offered infants by allo-mothers does not preclude assistance in situations where little harm or cost would be incurred by a volunteer. On three different occasions at Abu an infant who could not negotiate a leap was helped by a female other than its own mother. For example, when a four-month-old male (belonging to Elfin) hesitated before jumping from a tree to a nearby rooftop to join his mother, a middle-aged bystander (Earthmoll) picked him up and leapt with him, releasing the infant on the rooftop almost as soon as she landed. In a slightly different case, also involving Bazaar troop, a subadult female (Breva) who was carrying another female's infant halted at the tip of a branch and sat there, swaying in space, apparently afraid to make the leap carrying the extra burden of the infant. Moments later, a middle-aged female (Overcast) took the infant from its young caretaker and made the leap. Breva followed and retrieved the infant. Similar succor that benefits the recipient at little cost to the benefactor has been described for a number of other primate species (reviewed in Blaffer Hrdy, 1976).

Even though altruists are calculating individuals, it may frequently be beneficial to live in company with such relatives. Whenever the potential risk is low or the protectress is an old animal already approaching the end of her reproductive career, the potential gain in inclusive fitness promotes a certain amount of altruism towards nieces and nephews. And though they are not related to other adults in the troop, adult males traveling with the troop benefit them by protecting troop offspring, and benefit themselves, and incidentally others, by preventing marauding males from entering the harem. Aside from increased vigilance, the possibility of rescue, and a somewhat haphazard daycare service, there are other reasons why it behooves a langur to be social. In grooming, langurs present a combined front against ecto-parasites, cooperating to mutual advantage. By traveling together, information about food resources is shared. Typically one animal takes note of the discovery another makes simply by watching or else by approaching and sniffing the muzzle of a masticating fellow. Most importantly, traditions concerning the boundaries of the troop's range and the resources it contains must be transferred between female generations and this matrilineal legacy defended against neighboring lineages. Here again, older females benefit younger animals. When major shifts from feeding site to feeding site are made during the day, or when the troop leaves its sleeping site in the morning or heads towards it in the evening, such moves may occasionally be clearly attributable to a single animal who sets out ahead of the others and is later followed by them (fig. 4.8). In general, such moves are initiated either by the alpha male or by an older female.

The fact that most higher primates are profoundly gregarious creatures attests to the advantages, in terms of both individual and inclusive fitness, of living and cooperating with other animals, many of whom are relatives. Nevertheless, there are costs attached. Langurs living in groups may be more conspicuous and compete with others for resources. These costs will not be the same for every animal in the troop. Subordinate animals or immatures, for example, may suffer from competition more than an alpha male who can bully every other individual in the troop. Nor will the advantages be the same. Should a predator approach, an immature stands to gain from the experience and proximity of another animal more than a rather formidable adult male would. The current balance of these debits and credits is reflected in the degree to which some animals become peripheralized and become willing to leave the troop. As will be seen, there comes a time when even a creature so social as a monkey may decide to go it alone.

Fig. 4.8. The Toad Rock troop leaves its sleeping trees in the morning.

5 / Larder, Nursery, and Seraglio

In the midst of the perplexing changes that occurred between June 1971 and 1975, one feature of langur life remained remarkably stable: on any given day, a troop could be found within an approximately ninety-acre (0.38 square kilometer) overlapping plot. The boundaries of the home ranges represent cumulative occupation recorded during both wet and dry months, but because I was not present during all possible conditions, long-term ranges may be slightly larger.

Although the Abu home ranges appear small on the map, actual distances traveled by a troop over hilly terrain may be as much as 3.5 kilometers in a single day; shorter distances, however, are more common. Food availability and preferences, location of neighboring troops, location of suitable roosts to pass the night, as well as distance traveled the previous day, are the sorts of factors that probably influence daily routes, though a detailed study of route choices has never been made.

Each home range contains a number of sites suitable for langurs to spend the hours from dusk to dawn when their most dangerous predator, the leopard, is active. Typically, langurs will seek out well-developed groves of *Anogueissus* trees or silver oak (*Grevillea robusta*), or else high and expansive singleton trees such as banyan (*Ficus bengalensis*), Bengal plum (*Eugenia jambolana*), or mango

(*Mangifera indica*), all of which offer a measure of security. Roof-tops—usually those near trees—are also used. In all, some 35 different sleeping sites were recorded for the six troops living in and around Abu town. Of these, 27 were exclusively used by just one troop, while 9 other sites were regularly used by more than one troop (table 5.1). One large fig tree in the zone of overlap between the I.P.S., Hillside, and Bazaar troops was used by all three troops. Another exceptionally cosmopolitan sleeping site was the rooftop of an abandoned hotel called Lake House set in a mango grove on a peninsula jutting out into Nakhi lake. This site was used nearly every night by either the Arbuda Devi or I.P.S. troops, or by the various extratroop males who shadowed these two troops. On rare occasions, such shared sites are used concurrently. For example, during brief periods when Hillside troop was traveling without an adult male, the two troops sometimes spent the night in the same *Grevillea* grove only a few meters from each other. On other occasions, however, conflicts between two troops ensued when both arrived at the same sleeping site in the late afternoon.

Within each troop's home range, certain localities designated *core areas*, are used almost daily for feeding or sleeping, whereas other locations are visited more sporadically, once a week or less. Bazaar troop, for example, spends some part of almost every day scavenging in the bazaar. With the exception of nomadic males, who are not members of any permanently based troop, and of two Hill-

TABLE 5.1. Overnight sleeping sites used by seven langur troops in and around Abu town and at the Chippaberi bus stop, recorded between June 1971 and January 1974.

Troop	Known sites	Number of sites shared with one or more other troops
Bazaar	14	7
I.P.S.[a]	4	2
Toad Rock	9	3
Hillside	10	3
School[a]	6	3
Arbuda Devi[a]	3	1
Chippaberi bus stop	2	0

a. Because this troop was only occasionally encountered and followed, the record is incomplete.

side troop females who traveled temporarily as satellite members of Bazaar troop, no other langurs were ever seen in the bazaar.

Core areas tend to be in the interior of home ranges, but a border location may occasionally be the site of intensive use. In such cases, a core area may *not* be the exclusive domain of the residing troop, since trespassing by neighboring troops may occur in the course of intertroop conflicts. A focal point of the I.P.S. troop, for example, was an immense banyan fig that grew in the corner of their range adjacent to Hillside troop. In almost every respect, this was an advantageous habitat. The banyan grew next to the officers' quarters of the Indian Police Service Academy, and though occasionally harassed, the langurs were also frequently fed. In addition, the banyan was an ideal sleeping site: it was high and the nearby rooftop together with the extensive stalactite-like root system offered multiple escapeways. Because it was on the border of I.P.S. troop's home range, though, this banyan was vulnerable to incursions by the neighboring Hillside troop and was occasionally trespassed upon.

Though no quantitative data exist on this issue, food and security are probably the crucial determinants of core areas. Of fourteen areas of intensive usage, five contained locations such as temples, bus stops, or homes where the langurs were accustomed to being fed (fig. 5.1). Three were located near rubbish heaps or the bazaar where the langurs could depend on finding discarded (or else unguarded) chappatis grain, fruits, and vegetables. Almost all of these core areas contained trees or buildings suitable for overnight protection.

The most extreme example in which core area was synonymous with a preferred feeding and sleeping site is that of the troop at the Chippaberi rest stop, half-way between Abu road and Abu town. This troop spent more than 90 percent of its nights and part of almost every day around a banyan that shaded a vending stall and other outbuildings. Here pilgrims and tourists fed the langurs many times a day—whenever a bus passed through. The rest stop was surrounded on all sides by forest, and it is possible that the langurs also derived protection from predators by spending the night near human habitations.

Though multiple sleeping sites are clearly the rule among langur troops at Abu, there are exceptions, such as the Chippaberi troop. In general, those troops living in close association with humans appear to use fewer locations than troops in relatively undisturbed forest habitats such as Dharwar (where home ranges are smaller

Fig. 5.1. Mopsa, carrying her newborn infant, Brujo, takes food from a priest of Shiva who lives in one of the sacred caves or *gophas* in the hill-sides surrounding Mount Abu.

than at Abu) and at Orcha (where they are larger). As many as 25 distinct sites in 33 days were reported for one Dharwar troop (Sugiyama et al., 1965). Where resources are natural, transitory, and widely separated, and where no one location is more predator-free than another, such mobility makes sense. In fact, it may be safer for langurs to keep their whereabouts unknown to predators by constantly shifting their sleeping sites.

These observations suggest that nomadism within fixed home ranges is the normal pattern for langurs living in the wild. If so, the Chippaberi one-site pattern may represent an opportunistic adaptation to human bus schedules, while the pattern of the circum-town dwellers represents an intermediate compromise between the Chippaberi and Dharwar modes. If such a trend toward fewer sites does in fact exist, the main factors influencing it are likely to be the regularization and centralization of food sources and the reduction (though not the elimination) of predation. This simplistic hypothesis remains to be tested against the sleeping patterns of the many unhabituated forest troops who live on the hillsides surrounding Abu town, and who are not dependent on man.

The Seasonal Menu

Temperature and rainfall at Abu neatly fit into four seasons: the cold, dry months from December to February, when there is little precipitation other than haze and when night temperatures dip as low as four degrees centigrade; a moderately hot and very dry summer from March to about mid-June, when maximum temperatures reach 37 degrees; a temperate, wet monsoon arriving some time in June and lasting through August, during which Abu receives more than 95 percent of its annual rainfall; and a temperate postmonsoon season from September to November. Annual rainfall at Abu averages about 200 centimeters (or 80 inches). In a sample of 13 years taken between 1955 and 1975 (table 5.2), the minimum

TABLE 5.2. Annual rainfall on Mount Abu for thirteen years between 1955 and 1974.

Year	Rainfall (cm)	Year	Rainfall (cm)
1955	189	1961	313
1956	306	1962	121
1957	124	1970	210
1958	173	1971	134
1959	219	1972	106
1960	144	1973	359
		1974	39

Source: Courtesy of Soil Conservation Office, Dantiwara River Project of the Government of Rajasthan, and of Nirmal Kumar Dhadlal.

rainfall recorded (39 centimeters in 1974) was one-ninth the maximum (359 in 1973)—an indication of the extreme variability between years. These rainfall samples were taken near the top of Mount Abu at the same level as the town, where there is considerably more rainfall than at the base of the mount or at Chippaberi.

It is not possible to say with certainty that the dry season constitutes a season of scarcity. All langur weights were taken during the dry season, and no measurements were made with which to compare food intake in wet and dry seasons. From daily observations of the monkeys as they fed, groomed, and rested, the only detectable differences were the differing availabilities of, and the preferences for, certain foods, and the varying amount of difficulty and time spent in harvesting them. Despite the limitations of the data, I strongly suspect that the dry season occasionally is a time of serious food shortage. This opinion is based on the observation that rainfall is concentrated within a few months of the year and rainfall between years is extremely variable; the very desiccated appearance of the hillsides during the dry season even in a relatively wet year; and the fact that during the dry season the choice of food is limited, gathering it more difficult, and competition for advantageous feeding positions more frequent. During the dry season of 1975—the worst drought in Abu's recent history—the rate of naturally occurring displacements was approximately 0.4 displacements per hour that the langurs were watched at all. When the monkeys were actually foraging, the incidence climbed higher. On one day in 1975 I counted nine naturally occurring (that is, not provoked by human provisioning) displacements in 104 minutes, all over feeding positions that provided access to remnant pods hanging in a grove of Bauhinia trees. Unfortunately, I did not record the number of displacements per hour for the study periods prior to 1975, but it was my distinct impression that displacements over natural food sources during that time were less frequent (in fact, it was their rarity that initially discouraged me from collecting rates).

Their large diverticular stomachs allow langurs to subsist on quantities of mature leaves and to live in very dry areas. It has been reported that they can go for long periods without drinking water (Ripley, 1967b). At Abu, however, troops seek out natural and artificial water sources during the driest months of the year (fig. 5.2; cf. Beck and Tuttle, 1972). Because of water quests and the seasonal availability of certain foods, range usage varies slightly over the year. During the dry season Toad Rock troop made biweekly visits to an otherwise rarely visited small lake in the southwest corner of

Fig. 5.2. During the dry season, the langurs of Abu frequently make use of man-made water sources.

their home range, where they would drink and feed on the soft leaves of shore-growing plants such as the Indian willow (*Salix tetrasperma*).

During the driest months, shrubs (especially *Lantana camara* and *Carissa spinarum*) and trees (*Erythrina blakei* and *Bauhinia racemosa*) that provide staple foods for the langurs at other seasons become dry and lose their leaves. At this time of year, the langurs subsist on mature dry leaves, upon plants that grow year-round (such as *Anogueissus* and *Jacaranda* leaves and the cactus-like *Euphorbia*), and upon remnant seeds and pits of *Eugenia jambolana* and *Mangifera indica* left over from lusher seasons. During the dry season, the langurs make use of some plant phases that would be ignored by them at other times. For example, in December and January, tart green lantana berries are eaten, while during the monsoon season, only the very ripe, soft, black berries are taken. Similarly, though langurs apparently prefer the flowers and ripe berries of *Carissa spinarum*, in January the leaves are eaten as well.

Because not all plant phases are phenologically correlated, the dry season is not precisely synonymous with food scarcity. Months before the monsoon actually arrives, flush, flowers (on *Erythrina*, *Bauhinia*, and *Mangifera* trees), pods (on *Erythrina* trees), and immature fruit (on *Mangifera* trees and *lantana* bushes) are available. Several species of figs used by langurs (for example, *Ficus religiosa* and *F. bengalensis*), fruit on independent schedules, and figs on individual trees are available throughout the year.

In table 5.3, the multiple uses that langurs make of the same plant at different seasons of the year are diagrammed for five frequently used plants. A more complete list of staple and supplementary food plants used by langurs at Abu is provided in appendix 2.

The social organization of foraging among the related gray langur (*P. entellus thersites*) of Sri Lanka has been described by Suzanne Ripley (1970). As she points out, the epithet "leaf-eater" which traditionally has been applied to the various African and Asian members of the colobine subfamily is convenient, but in the case of *Presbytis entellus* somewhat misleading. The term implies a dependence on relatively abundant, well-dispersed food resources. In reality langurs are eclectic feeders who often prefer, and sometimes even depend on, such seasonal and finite items as fruits, seeds, and flowers. In the only quantitative study to date of langur food intake, langurs in Sri Lanka were observed for 118.8 hours and the estimated weight of everything they ate recorded (Hladik and Hladik, 1972). The diet of these langurs consisted of 21 percent mature leaves, 27 percent shoots (or young leaves and buds), 7 percent flowers, and 45 percent fruits. Given the importance of these seasonal resources and their limited quantities, langurs must sometimes compete for them.

It is extremely difficult in the case of any animal species to prove that food supply is the factor limiting population growth (Watson, 1970). Certainly casual observation of langurs does not immediately lead to this conclusion. When food is plentiful, langurs are notoriously wasteful and destructive feeders. After one of their visits to a jacaranda tree, for example, leaf-fringed branches will be strewn about the ground. Typically, langurs are transient browsers who move from any one feeding site long before they have exhausted all of the available food there. Could such happy-go-lucky feeders be food-limited? The answer is quite possibly yes. At issue here is the annual season of scarcity and the occasional drought year. Langur troops living in areas with regular provisioning by humans (as in the case of the Chippaberi bus stop troop or the advantageously situated I.P.S. troop) are far less subject to the vagaries of seasonal shortage. Not surprisingly, these groups were larger and faster-growing than are the more remote forest troops. Similar increases in troop growth rate among artificially provisioned free-ranging

TABLE 5.3. Multiple uses made by langurs at Abu of five plant staples.[a]

Date	Mangifera indica	Eugenia jambolana	Erythrina blakei	Ficus species (including 3–4 species)	Lantana camara
June 1971–72 and 1975	Green mangos			Ripe figs are available sporadically throughout the year	Flowers
1–10					
10–20					
20–30	Ripe mangos				
July 1971–72					
1–10			Flush		
10–20		Green jambols			
20–31		Ripe jambols[b]			
August 1971–72			Leaves		
1–10					
10–20					Ripe berries (ground feeding stations)
20–31					
September 1971–72					
1–10	Sprouts from germinated pits (on ground)	Dried remnant jambols (on ground)			
10–20					
20–30					

No information available from late September to mid-December

(continued)

TABLE 5.3. Multiple uses made by langurs at Abu of five plant staples.[a] (continued)

Date	Mangifera indica	Eugenia jambolana	Erythrina blakei	Ficus species (including 3-4 species)	Lantana camara
December 1973					
20-31		Flush	Bare	Green fruit	Green berries
January 1974					
1-10				Ripe fruit	
10-20				Leaves	
20-31					
		No information available for early February			
February 1973					
10-20	Small green fruit	Remnant jambols	Buds		
20-28					
March 1973					
1-10	Flush		Blossoms	Flush	
10-20				Green figs	
20-31					
April 1975					
10-20		Flush	Pods	Green and ripe fruit	
20-30					
May 1975					
1-10					
10-20					
20-31					

a. The information for each month was collected only during the years specified. Little data are available on variability of timing or of usage between years.

b. In 1971, the monsoon arrived early; the jambol fruits ripened by mid-July.

populations have been reported for Japanese macaques (Mizuhara, 1946; Iwamoto, 1974).

Because langurs are far from economical in their feeding habits, sizeable home ranges are a necessity. Continual exploration for perferred, seasonally abundant foods often leads langurs to impinge upon the home ranges of their neighbors. At both Polonnaruwa (Ripley, 1970) and Abu, langurs of one troop occasionally raid trees in the home range of a neighboring troop rather than feed on the same species within the confines of their home range boundaries. Such practices as itinerant feeding, sequential use of the entire home range, and constant exploration into the home ranges of neighboring troops are insurance against a time of shortage when a home range might represent the minimum area necessary for survival. For exactly the same reason, it is to both the short-term, and especially the long-range, advantage of the neighboring langur troops to police their larders against just such encroachments. Not surprisingly, conflict between troops sometimes ensues.

Encounters between Troops

During my first four fieldtrips to Mount Abu, hostile encounters between members belonging to different groups lasting anywhere from minutes to whole days were observed on seventy occasions. For simplicity, these encounters (itemized in Blaffer Hrdy, 1975) are divided into four general categories: (1) agonistic encounters in which a male (or males) from one troop temporarily leaves his own troop to visit another troop; (2) incursions ranging from temporary to semi-permanent by an extratroop male or male band into a bisexual troop; (3) encounters between two troops in which members of both sexes and various ages participate; and (4) encounters in which only females from different troops are engaged in hostilities.

Langur encounters involving whole troops (type 3 above) described by Sugiyama et al. (1965) and especially by Ripley (1967b), appear to fit classic definitions of territorial defense (Burt, 1943). Of twenty-six encounters between troops observed at Abu, 88 percent occurred in the zone where home ranges of the combatant troops overlapped. The pattern of conflict and the border locations of intertroop encounters observed at Abu closely parallel descriptions for Polonnaruwa and Dharwar. As Ripley has pointed out, troops often appear to seek each other out even though they are endowed with excellent intertroop spacing devices: the long-range male whoop vocalization; long-distance vision; and a predilection for high vantage points possessed by all males and at least some females.

Conflict between members of different troops may be occasioned by the immediate defense of a finite resource (for example, defense of a fruiting tree; the vicinity of human provisioning; an estrous female), but not all conflicts have such an immediately apparent incentive. It is tempting to interpet such aimless encounters as statements that indirectly reinforce a future claim, for example, over the minimum area necessary to survive a season of scarcity. Because so many encounters have no manifest intent, and because it is difficult to establish adequate controls and to quantify resources, the above speculations will be difficult to substantiate. Nevertheless, it would be of great interest to have data on *why* langur troops seek each other out and fight. Along with gibbons (Ellefson, 1968), titi monkeys (Mason, 1968), lutongs (Bernstein, 1968), and in some instances vervets (Struhsaker, 1967b), langurs are among the few documented cases of primates who regularly police their territories.

In most of the twenty-six cases of intertroop conflict witnessed at Abu, the two troops began drifting toward one another long before fighting actually broke out. As a result, it is difficult to pinpoint responsibility for the encounter; rarely could one troop be unambiguously assigned the aggressor's role. On only three occasions (in all of these Hillside troop invaded I.P.S. troop's favorite banyan sleeping site at the corner of their range next to Hillside troop's home range) did initial contact occur in what I considered to be a troop's core area. Usually the point of initial contact was in an area used by both troops. After initial contact, however, it was not unusual for an animal (especially a male) to trespass outside of his home range in aggressive pursuit of a member of the opposing troop.

Once the two troops were in close proximity, it was typically (in thirteen of twenty-six encounters) an alpha male who first actively displayed or actually entered the space occupied by the opposing troop. When his display or arrival triggers answering displays, chases, or even fights, it can be said that these males instigated the encounters. As other members of the instigator's troop moved up to join him, or as the invaded troop pushed forward to meet them, spats often broke out between other individuals.

A striking feature of many intertroop encounters is the alternation between hostility and casual coexistence. The capacity to ignore the close presence of a nearby troop may stem from long-standing familiarity between members of the two troops. For example, ten encounters were seen between the Toad Rock and School troops in about 300 hours of observation. In other words, these two troops convened and fought on average once every 30 daylight hours. Such encounters could last as long as 7 hours or more. This record makes individual recognition, at least by some members of the two troops, highly plausible. It was my impression that either because of the context or else due to his familiarity with his rival, Splitear (the Toad Rock troop alpha male) was far less agitated when he saw his next door neighbor Harelip (alpha male of the School troop) than when he espied an extratroop male.

Furious spats during which nothing appears to be either won or lost, interspersed with periods of adjacent foraging so calm that a monkey watcher on survey at that moment would record a single large troop with two adult males, typifies the relations between Toad Rock and School troops on occasions on which they met between June 1971 and February 1973, as the following long and highly repetitive account illustrates.

July 19, 1972: Between 8:15 and 9:30, the two troops leave their sleeping sites and drift toward each other. By 10:15, the two troops are 400 feet apart, apparently resting. At 10:20, Splitear, the Toad Rock troop alpha male, whoops, grunts, and leaps noisily about in the trees occupied by his troop. Two minutes later, Harelip displays. The two troops draw closer to each other. Indications of the rising level of excitement include chasing and mounting behavior among subadults, grunting, repeated embraces among adult females, as well as the alert stance taken by both males.

At 10:38, Harelip charges into the midst of the Toad Rock troop. Splitear chases him out. Both males withdraw. Harelip is out of sight. Splitear sits in the crotch of an *Erythrina* tree grinding his teeth. At 10:43, Harelip momentarily reappears. Splitear cackle-barks. Harelip climbs into a tree, grinds his teeth, and slaps his paw toward Splitear. Harelip is joined by an adult female and two juveniles from School troop. At this point, juveniles from the two troops begin to chase and wrestle.

At 10:45, Splitear moves toward a wall. The rest of Toad Rock troop is more or less gathered about him. Except for their tenseness and the screeching of juveniles, the scene is calm, until 10:54 when Harelip chases a Toad Rock juvenile without (apparently) making an effort to actually catch him. The adult male veers off from his pursuit and again climbs into a (different) tree. An adult female from Toad Rock troop chases two School troop juveniles.

At 11:00, Harelip lunges at a juvenile threateningly, but veers off at the last minute to take up a new perch on a large stone. An old woman approaches the langurs. She attempts ineffectually to drive them away. The two troops are approximately three meters apart, the two adult males facing one another, grinding their teeth. Suddenly, Splitear lunges at Harelip, who gives one whoop and runs away a distance of about a meter, then turns on Splitear. Splitear retreats, running back toward his troop. Harelip angles off from pursuit of Splitear and a Toad Rock juvenile, and ends up at the same rock where he began.

At 11:08, Harelip climbs onto a rooftop. At 11:11, Splitear follows him up; the two males whoop and chase each other about the roof. Both males withdraw.

At 11:45, 11:50, 12:20, and 12:25 the two males whoop by turns, after which they calm down. Between 12:25 and 1:35, the two troops doze and groom within meters of each other. At 1:55, Harelip resumes teeth-grinding. Splitear follows suit. Harelip leaps toward Splitear, who runs away, then turns and chases Harelip. Splitear returns to his tree. By 2:00 p.m., calm prevails. At 2:02, there are whoops in the distance; the two males grunt. Other members of their troop begin to stir, to forage and groom. Both troops feed on ripe jambol fruit in separate trees.

At 2:45, Harelip displaces a Toad Rock female. Splitear rushes to the scene and the two males threaten and lunge at each other. Harelip retreats.

Around 3:40, females from different troops enter the same *Inga dulcis* tree. As they slap at one another, Harelip joins the fray (3:55). He gnashes his teeth and threatens the Toad Rock females, at which point Splitear also enters the tree and chases Harelip. The two males briefly grapple before Splitear retreats.

A Toad Rock female accosts members of School troop. She advances a second time, engaging a female from the other troop in a slap-fight. A third time, she advances, and again the females slap at one another. The Toad Rock female retreats. At 4:05, Splitear chases a school troop juvenile. Other chases break out.

By 4:35, the two troops relax again. All forage except for the two males, both of whom survey the scene tensely. Immatures from both troops play derring-do with adults of the other. Occasionally, a mother rushes to retrieve her infant. By 4:40, the two troops have spread apart so that there is a corridor of no-langurs' land between them. In addition to this buffer, the two males avoid looking at each other.

At 4:55, an old woman and her children come to gather wood under the large mango tree occupied by Toad Rock troop. Something (the antics of the juveniles?) precipitates renewed hostilities. Splitear leaps right over the heads of the humans, landing beside Harelip, who may have been using the people as cover to approach. Harelip chases a Toad Rock female momentarily before veering off to reassume his former perch.

Accompanied by much grunting, another chase erupts in the mango tree occupied by Toad Rock troop (at 5:07). Splitear cackle-barks and gnashes his teeth. At 5:15, Harelip again approaches the base of the mango. Nearby juveniles squeal, but a Toad Rock female about three meters from Harelip sits quietly, blasé about the intrusion. Splitear is out of sight on the other side of the tree. At 5:30, Harelip makes an uneventful withdrawal. Other troop members are busy eating *Erythrina* leaves.

At 5:35, Harelip approaches the *Erythrinas* occupied by Toad Rock troop. Splitear chases him. A School troop juvenile also approaches and is chased back. Two minutes later, Splitear follows Harelip, chasing him right past his own troop until Harelip turns and chases Splitear back to the Toad Rock-occupied mango. On arriving, Splitear whoops and displays.

Within fifteen minutes of the start of this chase, the two males draw apart. Splitear sits on a granite outcrop overlooking the route the two troops took that morning. Occasionally Splitear grunts. Toad Rock troop drifts back toward Jaipur House, arriving at that core area sleeping site by 7:30 p.m. Harelip's troop has withdrawn into the vicinity of its core area at Sophia School.

After more than seven hours, the two troops separated and returned to their original starting points. Despite the expenditure of much energy by almost every troop member, there was no immediate gain to either side. In contrast to the highly goal-oriented aggressiveness exhibited by langurs during male invasions, combat during intertroop encounters was highly stylized and was never observed to result in injury. Of the four categories of intergroup conflict, these border encounters involved the broadest spectrum of troop membership.

Though it is not yet possible to conclusively show what langurs are defending when they fight, it is assumed here that conflict between groups reflects competition over prerequisites of survival and reproduction. Such prerequisites, of course, differ according to the sex, age, and current reproductive role of the individuals concerned. But in general, members of the same troop share a stake in access to safe sleeping sites and adequate foraging areas, as well as in the protection of immatures against marauding conspecifics. In addition, adult males stand to gain in fitness from maintaining their (more or less) exclusive sexual access to fertile females in their troops.

Encounters are classified as 3 ("intertroop") if any troop members in addition to adult males are involved. Hence, if a male chasing another male veered off and momentarily chased a female, or if a female lunged at a male who had accidentally threatened her infant, the incident was counted in the intertroop sample of twenty-six. However, on at least fifteen occasions (58 percent of the time), females played major offensive as well as defensive roles. On five of these occasions, it was a female rather than an alpha male who first instigated fighting. Only once, however, was a female (Pawless of Hillside troop against Bazaar troop in 1973) the first to actually approach the other troop.

Generally, the most aggressive females are those not currently suckling an infant, though a mother sometimes left her infant with or near another female on the periphery of the action while she participated. On rare occastions, mothers entered the fray with dark or cream infants clinging to them (fig. 5.3), but the mother of a very new baby was never seen to do this. Likewise, pregnant mothers on the eve of delivery may hold back from combat.

Female hostility toward outsiders is manifested by chasing or by facing an opponent, grimacing, giving open-mouthed threats, lunging, and slapping (fig. 5.4). Juvenile males and older infants also participate in encounters, usually as charging, squealing *agents provocateurs*, challenging members of the opposing troop—especially adult males. Such derring-do rarely gets the immatures them-

Fig. 5.3. An I.P.S. mother with newborn confronts a triumvirate of Bazaar troop females (including Breva, the short-tailed female in the rear) in the overlap zone between their two home ranges.

Fig. 5.4. Lunging in place, fierce slaps, and a frightening grimace are typical behaviors for females during intertroop encounters.

selves into trouble, but it may bring into the fight adult females who rush to protect them (fig. 5.5). Chases and wrestling bouts among older infants and juveniles from different troops were not distinct in appearance from ordinary intratroop "play" behavior.

The participation of subadult and young males varies according to the situation. In some cases young adult males from the troop (for example, the two young males in Bazaar troop in 1971) participated along with the alpha male in chasing away intruding males. Similarly, during the period when five extratroop males had joined Mug in Hillside troop in 1973, all six males would advance in unison to confront Bazaar troop. And if the alpha male from Bazaar troop advanced, all six fled together. In yet a third instance of young males cooperating in defense, eight juvenile-to-subadult males who had been ousted from the Toad Rock troop, together with the former alpha, Splitear, jointly continued to patrol the Toad Rock home range following their ouster. Together, the eight young males and the former alpha attacked an encroaching male band. Next to Splitear himself, the oldest three subadults were the most active participants. On three occasions, these three young males made raids on their own, independently of Splitear, into School troop. Each time, they copulated with females in the troop.

Despite this continued involvement in territorial defense after their ouster, none of these eight had participated noticably in troop defense in the years prior to it. Furthermore, in 1971 and 1972 when

Fig. 5.5. Two parous females, one with an infant, spar at the interface between their home ranges.

there happened to be two older, nearly adult males (Beny and No. 2) in the troop, neither had participated in intertroop encounters— even in 1972 when Beny was almost an adult and at least as old as the youngest Bazaar troop male had been in 1971 when he assisted in troop defense. In contrast to the calm that generally prevailed in the relations between the three Bazaar troop males, the relations between Beny and the Toad Rock alpha male, Splitear, were tense. The antagonism between them was especially manifest when there was an estrous female present in the Toad Rock troop. Splitear would threaten and chase Beny if he approached her. When I recontacted the Toad Rock troop in February of 1973, Beny and a slightly younger male had disappeared. From the above examples, it appears that young adult and older subadult males do participate in troop defense under some circumstances, but that they do not necessarily cooperate with the alpha male in routing intruders (see Poirier, 1970, for *Presbytis johnii*). The subadults in turn may invade other troops on their own, without the aid of the alpha male.

This small sample of twenty-six encounters suggests important differences between male and female roles in intertroop conflicts. The majority of encounters were initiated by the alpha males; every encounter eventually entailed male threats. Such threats could be confined to vocalizations (whoops; grunts accompanied by biting air; and occasionally cackle-barks) or else could erupt into opportunistic displays such as ricocheting off branches or water pipes and thrashing about in the trees. The combination of these antics made for an impressive audio-visual effect before any rival troop. During twelve of the twenty-six encounters, males chased each other at least once and usually more. Despite the frequency of threats and chases, physical contact between males (that is, catching hold of one another, grappling, and biting) occurred on only five occasions, or 19 percent of the time, and usually terminated within seconds as both males withdrew. Lunges in place while grimacing or slapping fiercely (color fig. 3), abortive chases, and dramatic leaps that terminate harmlessly, are by far more typical. Color figure 4, for example, depicts a spread-eagled, open-mouthed Splitear hurtling dangerously through space. In fact, he landed nowhere near Harelip. By contrast, females chased other animals less frequently (in only eight of the twenty-six encounters), but they contacted each other (by lunging and slapping at close range) relatively more often than males did (46 percent of the time).

All-out fights between males are rare, probably because of the formidability of another male langur and the diminution in reproductive potential that a male would suffer if injured. Nevertheless, threatened attacks are somtimes carried out. Whether a male fights or not perhaps reflects the balance between the obvious disadvantages of fighting and the losses entailed by *not* fighting—as well as the possible benefits to be gained from winning. The tally will differ according to the size and physical condition of the individual and his current reproductive situation (that is, whether he is with or without a harem; whether any females in his troop are currently ovulating; the number of reproductively active years that remain to him).

Relative to extratroop males, alpha males are in an advantageous position insofar as they control reproductive access to females in their own troop. The gain to an alpha male for fighting would usually be one of degree—that is, he might gain access to a temporary food source or access to more fertile females than are currently available in his own troop. The loss to him if injured, however, might be absolute if he subsequently lost his harem due to injury-related debilitation. His existing offspring would be killed by invaders and his future access to fertile females preempted. Futhermore, alpha males very rarely gain long-term benefits from taking over additional troops. Even when an alpha male was able to usurp his neighbor's harem, in four of five cases he was unable to retain control over both at once (Sugiyama, 1966; Blaffer Hrdy, 1977a) In most cases, antagonism between females in the different harems and their resistance to any merger meant that eventually the double-usurper had to choose between his two troops. Only in one case (that of the 2nd and 3rd troop merger at Dharwar) did a double-usurper find himself in possession of a merged troop (Sugiyama, 1966). In other words, taking over a second troop is not normally a profitable alternative, although when the gamble does succeed, the pay-off might be potentially great.[1]

The cost and benefits attached to fighting might be quite different, then, for those males already in possession of a harem and those males without one. While in most cases harem alphas have more to lose than an invader, this might not be the case if a young, newly entered alpha male were pitted against an old invader approaching the end of his reproductive career. Whether or not alpha

---

1. In two cases, that of the 4th troop at Dharwar (Sugiyama, 1966) and Shifty's take-over of Hillside troop at Abu, the double-usurpers probably sired offspring in both troops before giving up one of the two. Only in the case of the 4th troop's usurper, however, did offspring in the "deserted" troop survive. All the infants in Hillside troop that were presumed to have been sired by Shifty (Mira, Scratch, and possibly Pawla) were killed (see chapter 8).

males have as much or more to gain than the invader could depend on such factors as the average remaining tenure of the alpha male, the number of still vulnerable offspring he might have in the troop, and the probability that he will retake this or another troop. In general, however, nomadic males have a great deal to gain and not such a great deal to lose by invading, unless they receive an injury so serious that it precludes future reproductive activity. Thus, even in the case where one male is much more powerful than his opponent, aggressiveness and willingness to invest in the outcome of a conflict may vary, and the resolution of such competitive situations will not always be clearcut (Maynard-Smith and Price, 1973; Parker, 1974; Popp and DeVore, in preparation).

The reluctance of males to engage in physical combat and their reliance instead on bravado may be compared to nuclear powers held in check by the prospect of retaliation. For an alpha male possessing a harem in the competitive world of Abu, any injury that leads to even temporary debilitation may mean the loss of his harem to another male: no male in less than excellent physical condition was ever seen in command of a troop. And any fight with another powerfully equipped adult male always entails the risk of reproductive wipe-out because of serious injury.[2]

Females, on the other hand, less muscled and weighing on average some 15 pounds less than males and possessing less dangerous canines, run a smaller risk of serious injury from fighting among themselves. The evidence on langur pathologies indicates that adult males are far more frequently injured in fights than adult females are. Whereas 39 of 48 (81 percent of) adult males showed signs of having been injured, only 23 of 70 (33 percent of) females did. And when injury does occur among females, the price they must pay for it is not as high as the price paid by an alpha male. Several females at Abu suffered from serious disabilities. Pawless in Hillside troop, one-eyed Wolf, Overcast, armless Guaca, all of Bazaar troop, are cases in point. Even though the low rank of these females in the female hierarchy may have been at least partially

---

2. The view that all-out attacks on conspecifics are forestalled by the possibility of retaliation is consistent with the results of a rather curious experiment at the Oregon Regional Primate Center. In order to prevent injury either to the monkeys or to their keepers, the dagger-like canine teeth of seven adult Japanese macaque males were extracted. The position of high-ranking males was not noticeably affected by the loss of this weaponry, but low-ranking males without canines, males who had lost the potential of damaging retaliation, were subsequently attacked and killed (Alexander and Hughes, 1971).

imposed by their physical disabilities, these females participated in all phases of troop life. Pawless, for example, was one of the most aggressive females in intertroop encounters and in defending Hillside troop infants from the attacks of infanticidal males. To the extent that rank is correlated with fitness and infant-care is hampered by their disabilities (especially the case of primiparous Guaca in 1974), these females may leave fewer surviving offspring than their more hale and hearty comrades; all of them, however, have borne at least one offspring.

At Abu, the most earnest fights witnessed were between harem alphas and nomadic males. Similarly, Yoshiba has reported that at Dharwar, "the relation between the bisexual troop and the all-male group is more aggressive than that between bisexual troops" (1968; see also Sugiyama, 1967). One way of interpreting this antagonism is that the encounter involves a male with an investment confronted by a male who has less to lose and much to gain and who, for this reason, presents a serious threat.

By contrast, when two alphas meet (as in the case of Harelip and Shifty) each stands to lose a great deal if wounded and to gain only a relative amount if successful. Because of this disadvantageous balance between pay-off and risk, both alpha males would be under similar constraints; neither represents so great a threat to each other as a nomad does. This feature of their relationship could explain why the proximity of other alpha males is often tolerated in a way that the proximity of a strange nomad almost never is, as well as why serious fights between alpha males were so rare at Abu. Similarly, Sugiyama has written that at Dharwar, the invasion of a neighboring alpha male into a troop "rarely develops into a severe biting fight between the two males" (1965:95).

These explanations are based on the assumption that combatants are more or less equally matched. However, two males or two troops are not always equal in size and in readiness to fight; one party may consistently retreat before the other. Furthermore, such relations between individuals and groups are not stable over time.

Encounters between the Toad Rock and School troops between 1971 and February 1973 were characterized by an equilibrium in which members of both troops avoided all-out fights, and neither troop pushed the other very far. During this period, neither alpha male was ever observed in the core area of the other troop. The first time such a core area intrusion was ever observed was on December 23, 1973, at the time of drastic changes in the internal politics of Toad Rock troop. During the period prior to December, Splitear, the long-time alpha male of Toad Rock troop, was ousted by a new male, Toad. Together with eight juvenile-to-subadult males, and for brief periods three Toad Rock mothers and their offspring (referred to here as Splitear's contingent), Splitear continued to forage, to sleep in, and even to police the Toad Rock home range against intrusions by the Waterhouse band of males. Between Splitear and his former troop, there was a policy of mutual avoidance. Despite the fact that it was highly unusual for two bisexual troops at Abu to occupy adjacent (not to mention the same) ranges without one or both seeking the other out, in a period of 144 hours when I was watching one or the other, the two groups met only once. On this occasion, Toad and his new harem immediately withdrew.

Toad and his new troop may also have been avoiding School troop. Whereas previously the two troops had met every few days, during the 1973-74 study period just after Toad took over, they were never once observed to meet. [3] Though Toad and his troop apparently steered clear of the School troop boundary, Splitear and company appeared to linger there. On seven of nine days that Splitear's contingent was observed, they were in the zone of overlap between the School and Toad Rock troop home ranges for at least part of the time.

In addition to lingering at the boundary of School troop, Splitear also appeared more aggressive toward his neighbors. Food tests administered casually both before and after Splitear's ouster from his troop support this interpretation. Two food tests between School and Toad Rock troops were made in 1972 while Splitear was still the alpha male of the Toad Rock troop. On July 4, 1972, the two troops were arrayed against each other on either side of a stone wall: Toad Rock troop above the wall, School troop below it. Occasionally, troop members would cross the wall, spat briefly, and return. Otherwise, the wall proved to be an effective boundary between them. Even when chick-peas were thrown right over the heads of Toad Rock troop to School troop members, Toad Rock troop could not be induced to cross the wall, and vice versa. Only after all feeding had stopped and the School troop had moved away uphill did Toad Rock troop (30 minutes later) cross the wall and follow School troop up the hill. On July 6, 1972, the two troops met again in the

3. By April 1975, the old pattern of encounters between these neighbors had resumed. During the 1975 dry season, Toad Rock and School troops met approximately every other day. As before, relations between Harelip and the Toad Rock alpha male ranged from mild antagonism to actual fights.

overlap zone between their ranges. When food was scattered between them, School troop took over the site while Toad Rock troop withdrew.

Seventeen months later, however, after Splitear had lost control of his troop, a new food test elicited a more aggressive response from Splitear and his contingent. At noon on December 23, Harelip approached Splitear, the eight other ousted males, and two mothers. Together, Splitear and Pawlet chased Harelip away. For the next two and one-half hours, chases and brief spats broke out between the two groups, until School troop retreated into their core area at Sophia School. What happened next was unprecedented. Splitear and his contingent pursued Harelip and his troop further into their own core area. When food was scattered between the two groups, fights involving both males and females broke out between the two groups. After feeding stopped, School troop retreated deeper into their core area. At 3:30 p.m., Splitear's group followed them. Observations of the encounter terminated at 3:40 when a sub-adult female from the Bazaar troop stole an infant from School troop.

There are several possible explanations for Splitear's increased aggressiveness, among them the curtailment of ranging space due to the pattern of avoidance between the ousted males and the newly formed Toad Rock troop, and the possibility that having lost his harem (soon after the December 23 encounter, even Pawlet and P-M defected, taking their infants with them), Splitear had less to lose by fighting. Whereas Splitear's constraints against fighting were diminished, Harelip's were not.

The case of Shifty versus Mug is more perplexing. When first encountered in 1971, Mug was a male in his prime, heavyset and impressive. Shifty was definitely a smaller male: his sunken eyes, tattered ear, and the missing scruff of hair from his sideburns gave him a worn look suggesting that he was also substantially older than Mug. Despite his smaller size, Shifty was able to usurp Mug's troop in 1971. Furthermore, even after Shifty left the usurped Hillside troop for the larger Bazaar troop in the spring of 1972, Shifty continued to drive Mug away from Hillside troop whenever the former leader attempted to reinstate himself. On at least four occasions in 1972 Shifty chased Mug out of Hillside troop. Even after Mug was joined by five additional males during the early months of 1973, Shifty was able to chase out all six males on five occasions.

The consistency with which Mug fled from a confrontation with Shifty may reflect differences in the experience or physique or in the "personality" of the two males which rendered Shifty Leftless more powerful. If Shifty was in fact older than Mug, however, a strong case could also be made that Shifty had a lower reproductive value than Mug did and that Shifty's greater aggressiveness was a component of a now-or-never policy.[4]

With the passage of time, however, Mug's position with respect to Shifty seemed to improve. By December and January 1973-74, Mug no longer ran away from his troop whenever Shifty approached. In three encounters between the Hillside and Bazaar troops during this period, Hillside troop retreated, but Mug remained with them. On a fourth occasion (Christmas day, 1973), Shifty was temporarily absent from the Bazaar troop, and the Hillside troop was able to prevail in a food test between the two troops. In another food test between the two troops with Shifty present, Mug at first avoided Shifty by hiding behind his females, but then emerged and actually grappled with him momentarily before withdrawing. One variable that may have deterred Shifty from leaving Bazaar troop to chase Mug as he had in the past was that on all three occasions one of the females in his troop was in estrus.

The events observed between 1971 and 1974 suggest a trend of increasing boldness on Mug's part. When I returned in April of 1975, Shifty Leftless had disappeared from Bazaar troop. In his place was the newly intrepid Mug.

Though on the surface size would seem to be an explanation for Hillside troop's vulnerability, the full array of intertroop incidents does not support this interpretation. In the following episode, which was witnessed at the beginning of the study, Hillside troop was able to rout the much larger Bazaar troop (at that time containing three resident adult males) first from a lucrative feeding position in the dump and then from a chosen *Anogueissus* roosting site.

> July 11, 1971: Bazaar troop is scavenging in a refuse heap beside Rajendra Road, at the border of their home range. The Bazaar troop alpha male climbs to a high vantage point on a rooftop and stares to the southeast (toward Hillside troop's core area near Hillside house). He grows increasingly agitated, throws back his head and utters a

---

4. According to Fisher (1958), "reproductive value" refers to the extent to which individuals of a given age will on the average contribute to the ancestry of future generations. Because an older primate will have fewer remaining years for reproduction, invariably reproductive valve would be lower than for an animal just beginning to reproduce.

series of low resonant whoops which give way to a crescendo of much faster whoops.

In the trees of Hillside yard, across Rajendra Road more than 100 meters away, three Hillside mothers are visible. At 4:50 p.m. Mug comes out of the forest near Hillside house and lopes purposefully across Rajendra Road to the dump. When he is about 3 meters from the three Bazaar troop males, he whirls and runs back to his own troop who are advancing to join him. Mug runs past the approaching females into the forest, with the three Bazaar troop males in pursuit after him. As the three males fan out in their chase, Mug turns again and chases the third-ranking male back to his troop. As he gains on him, Mug veers aside and pauses.

The three Bazaar troop males take this opportunity to regroup; they circle about one another "conversing" in low grunts. Meanwhile, Pawless and Sancho appear on the scene. With Pawless' support, Mug makes a second foray into the Bazaar troop, and then retreats to where other Hillside females wait on the dump side of Rajendra Road.

Moving separately, the two troops travel uphill into an area often frequented by the Bazaar troop, but only occasionally used by Hillside troop. The Bazaar troop groups together and climbs into a small *Anogueissus* tree where they feed til six o'clock, when Mug, followed by Pawless, suddenly charges the tree. In the ensuing commotion, Mug and Pawless vigorously chase Bazaar troop members. Despite the unequal numbers (two from a troop with seven adults versus a much larger troop with thirteen adult and subadult members), when it is over, Mug sits by himself in the *Anogueissus* tree. Gradually, the rest of Hillside troop joins him there. The Bazaar troop is grouped compactly on a flat rock 10 meters from the base of the tree. Even while fighting is going on, juveniles and infants from the two troops engage in play and play-fighting. After the fighting ends, they continue to chase and wrestle in the space between the two troops.

From these events, it appeared that Mug had the upper hand. Nevertheless, not ten minutes after their victory at the *Anogueissus* tree, Hillside troop abandoned the prize:

At 6:20 p.m. Hillside troop leaves the tree, at which point a subadult female and some juveniles from Bazaar troop reclimb it. Noting this incursion, a Hillside troop mother deposits her infant with another female and rushes back up the tree. The interlopers hastily scurry back down the trunk.

Once again, both troops travel separately toward the northeast. Bazaar troop, followed by Hillside troop, moves into the mango grove behind the National Police Academy (a shared sleeping site of both troops). Hillside troop appears to be settling down for the night when a Hillside infant scampers down the tree the troop was in, leaps onto a wall that runs between the two areas occupied by the troops; single-handedly the squealing infant charges the entire Bazaar troop. Two adult and one subadult females from Bazaar troop respond by rushing at the tiny challenger. Pawless runs to the infant's defense, engaging the three females in the fiercest slapping encounter of the evening. Apparently reluctantly, the adult males are drawn back into the action. Males whoop and display. When an alpha male from Bazaar troop lunges at a Hillside mother and infant, the skirmish is escalated into a face-to-face grappling match between the Bazaar troop alpha male and Mug. The duel ends when both males turn and flee at the approach of three large dogs. The two antagonists stare at each other from separate tree perches, but do not appear eager to resume fighting. For a time, both troops watch their common enemies barking below, then Bazaar troop retreats into the heart of its home range to spend the night. Hillside troop sleeps where it is.

I have chosen examples that illustrate how highly conditional superiority in intertroop or intermale encounters may be. Such factors as location of the conflict, the reproductive state of females in the troop (including estrus, pregnancy, and the presence of very young infants), the fighting ability of males, as well as their willingness to fight, may enter into the balance. Troop size may also be a factor (Jay, 1965), but it is unlikely that this is the only or the crucial factor.

The Pros and Cons of Male Invasion Strategies
Nomadic males crisscross the home ranges of a number of bisexual troops. When they meet, relations between these wandering males and troop alpha males are almost invariably hostile. All such encounters observed at Abu were initiated by extratroop males. These male invasions may take any one, or several sequentially, of the following forms: "haunting"; attacks (leading to chases and combat) ending in retreat; temporary joining, in which both resident and invading males remain; and take-overs in which resident males are driven out.

A single nomad, male band, or males from an adjacent troop may haunt a troop by seeking it out and by spending periods of time on its periphery. On more than sixteen occasions observed at Abu, such haunters were chased away by resident males from the Chippaberi, I.P.S., Toad Rock, Hillside, and School troops. Nevertheless, the trespassers often returned to skulk at a just-tolerable distance from the troop. The following example taken from field notes illustrates the single-minded preoccupation and persistence sometimes shown by haunters:

> At 6:50 on the morning of August 19, 1971, two nomads stir at their sleeping site in the mango grove beside the abandoned Lake House hotel. Their first movement of the day is to climb to a high vantage point on the hotel's roof and to survey the area. Soon after, both males head eastward to the precise spot where I.P.S. troop spent the previous night. I.P.S. troop has already moved away, and it is 8:35 before the two males catch up with the I.P.S. troop. As soon as LeGrand detects them, the I.P.S. troop's resident male chases them away, but the two males remain in the vicinity of the troop throughout the morning.

Occasionally, instead of retreating before the resident male, extratroop males stand their ground in the face of his assaults, and may even chase him. Eventually, such males may be fought off or leave of their own accord. For example, on May 31, 1962, a band of seven males invaded the 30th troop at Dharwar and severely wounded the resident male, Z. Violent counterattacks were made by females against the invaders. Within hours of their entrance, all seven males retreated. On the following day, the males again attacked, and again retreated (Sugiyama, 1965b).

On several occasions, instead of being chased out, alpha males and invaders coexisted. During February and March 1973, five males in addition to the resident male, Mug, traveled with the Hillside troop for some part of 29 of 41 days on which the troop was observed. Similarly, between two and five extra adult males were present in the Chippaberi troop on the 11 days between August 21 and September 13 when this troop was observed. It was suspected that these additional males were coexisting with the resident male, Big Plain, but because this alpha male had no absolutely identifiable marks, this suspicion could not be confirmed. Similar male sojourns in bisexual troops were recorded by Vogel (1973a) at Sariska, by Bishop (1975b) at Melemchi, and by Boggess (1976) at Solu Khumbu.

If they were able to, invading males would drive off the alpha male and usurp his troop. For example, on the third day following the May 3, 1962, attack at Dharwar described above, the male band once more invaded the 30th troop. The male Z again fought with the intruders and injured his major opponent, L, but in the process suffered a serious injury himself. After this battle, the severely wounded and bleeding Z was chased away from his troop by L. On the fourth day, the other six members of the invading band were evicted by L, who by this time had worked his way into the position of 30th troop's new alpha male (Sugiyama, 1965b). As is typical of such take-overs, the evicted males remained for some time in the vicinity of the troop and made several counterattacks.

These male invasions differ substantially from the intertroop encounters described earlier in this chapter. Whereas 88 percent of encounters between troops occurred at the borders where home ranges of the two troops overlapped, males invaded troops wherever they happened to be, even in their core areas. Seventy-six percent of the encroachments by nomadic males and 36 percent of those by temporary philanderers from neighboring bisexual troops occurred in the core areas of the troop invaded. In all three types of encounters, resident alpha males played prominent roles in troop defense, but in the case of intertroop encounters other troop members, including females and immatures, also participated and occasionally even initiated hostilities. By contrast, when alien males skulked about on the outskirts of troops, neither females nor immatures threatened or chased them; in general, defense against haunters was a strictly male affair.

Females could also be relatively tolerant of alien males traveling temporarily with the troop. [5] During the many occasions when alien males foraged, dozed, or groomed in the midst of Hillside and Chippaberi troop members, females were seen to threaten or attack them in only two contexts: when such males happened to threaten an infant or when invaders attempted to copulate with another female troop member. But female tolerance toward males traveling with the troop does not extend toward males taking over a troop. In one of the earliest records of a take-over, Hughes (1884) reported that females fatally wounded one of the invading males. Sugiyama (1965b) described furious attacks by females against the band of seven males who invaded the 30th troop at Dharwar. Similarly, Hillside females chased and harassed Shifty in the first weeks after his take-over in 1971. Whenever Mug (in 1972) or Righty (in 1975) attacked infants in the Hillside and Bazaar troops, females responded with furious counterattacks. In each of these cases, female antagonism was justified by the usurping males' very real threat to survival of troop infants.

5. Obviously, another explanation besides tolerance would be fear of the alien male, who was after all dominant to most if not all females in the troop. Nevertheless, females can and do challenge alpha males in other troops in the course of intertroop conflict.

The most significant exceptions to the female policy of avoidance and antagonism were exhibited by estrous females. At all sites where relations between females and invaders have been described (Dharwar, Jodhpur, and Abu), estrous females sought out and solicited both invaders and usurpers. In fact, there is some indication that the presence of alien males may be a stimulus that induces estrous behavior in some females. It is important to note the distinction made here between estrous behavior, which can be observationally detected, and ovulation, which cannot, since the two do not necessarily coincide (Loy, 1970). There is evidence that pregnant females exhibit estrous behavior and that they may be most likely to do so in the presence of an alien male. Sexual solicitations of neighboring alpha males are not uncommon at Abu. A female may solicit neighboring alpha males from a distance while the two troops happen to be in proximity; she may solicit a neighboring male who has entered her troop and first approached her; or she may approach the male herself prior to soliciting him. S. M. Mohnot has observed similar behavior among females at Jodhpur (personal communication). In each instance where observed solicitations might have led to further contact between the female and a neighboring alpha male, the female's own alpha male intervened, either by herding the female back into her troop or by chasing out the trespassing male.

No solicitation of neighboring alpha males (except in the case of double-usurpers) ever culminated in the female being mounted. Nomadic males haunting troops did not fare much better. The only exceptions that I know of occurred when an alien male, Newcomer, entered the Chippaberi troop and when, on two successive occasions, subadult males recently ousted from their own Toad Rock troop (and at that time traveling with the former alpha male, other ousted immatures, and three mothers) were able to sneak into the School troop and steal copulations at times when Harelip was temporarily absent from his troop. The circumstances surrounding these last two raids are rewritten from the field observations and 16 millimeter film record of D. B. Hrdy for December 12, 1973, and from my fieldnotes for January 1, 1974.

(1) On the afternoon of December 29, 1973, Splitear's portion of Toad Rock troop, and the School troop both forage near Windermere Lake. Three Toad Rock subadults (Cast-eye, Blind-ear,[6] and No. 3)

6. On December 28, 1973, it was noticed that this subadult's left ear had been torn off, almost surely in an attack from an adult male langur. Sugiyama observed

approach the School troop. Harelip chases two of them back to where Splitear's group is foraging, in the cemetery, approximately 0.5 kilometers away. While Harelip is gone, one of the Toad Rock sub-adult males enters School troop and copulates with a young estrous female during two bouts of mounting and thrusting which take place at about 15-minute intervals. Approximately 30 minutes later, Harelip returns and chases out the third subadult.

(2) In the early afternoon of January 1, 1974, the School troop is resting in a *Ficus* tree growing up from the canyon floor seven meters below, and on a rounded granite outcrop level with the lower branches of the tree. Thirty meters further down the canyon (to the northwest), Splitear's remnant of the Toad Rock troop dozes in a large *Mangifera* tree. A house (Shiv Kuti) and one wall of the canyon connect the area between the two troops.

At 1:35 p.m., Harelip in the company of two middle-aged multi-parous females follows a path descending down from the outcrop, toward the southwest. After his departure, at 1:57 a young (probably primiparous) estrous female turns her rump toward the Toad Rock group and shudders her head. She looks over her shoulder at the oldest Toad Rock subadult male (Cast-eye), who has an erection.

At two o'clock, Cast-eye, followed by the next oldest subadults (Blind-ear and No. 3), crosses the land bridge to the soliciting female. She continues to present and shake her head as Cast-eye mounts her. A second subadult nuzzles her breast and appears to be trying to suckle,[7] though this female was not known to be lactating. A third subadult mounts the female, but does so incorrectly, and thrusts at her back. School troop juveniles approach and ineffectually harass both the female and the intruding subadults; temporarily, one of the subadults is able to threaten them away.

While two of the subadults slap and lunge at the harassers, the third copulates. At 2:03, scarcely six minutes since the invaders entered his troop, Harelip returns and breaks up the consortium. The three subadults return abruptly to the *Mangifera* tree; Harelip chases the males briefly, then returns and sits next to the young estrous female. She solicits Harelip, but he ignores her.

Any nomad temporarily able to join a troop has a better prognosis for successfully copulating with troop females than a haunter does. Despite the fact that invading males in both the Chippaberi and Hill-

---

precisely the same injury inflicted by one langur male on another in a fight at Dharwar (1965b:392). It is possible that events of December 29 and January 1 may have also occurred earlier as well. If so, Harelip would be a possible suspect for the male who attacked Blind-ear, but we were not able to confirm either suspicion.

7. Struhsaker (1975) reports two cases in which adult male red colobus monkeys suckle from adult females; I know of no other parallels for this peculiar behavior.

side troops competed among themselves for access to troop females, and even though mounted couples were harassed by other invaders and troop members, the odds favor the eventual (if limited) success of any male able to permeate a troop. Some seventy adulterous solicitations (solicitations of outsiders by troop females) followed by mounting behavior were observed during temporary invasions of these two troops. Of six males—at least five of which were invaders —that were observed in the Chippaberi troop in August 1971, all had at least one opportunity to copulate with an estrous female. Approximately 83 percent of 18 mounts observed were harassed by other invaders or by troop members, but on at least 8 occasions, males achieved intromission. Five of the six males present could be ranked in order of their ability to displace another male over access to either food or position (Scarface, Big Plain, Misshapen, Left-lip-scar, Jag). This displacement did not exactly coincide with body size order from largest to smallest (Misshapen, Scarface, Big Plain, Left-lip-scar, Jag). There was little indication from this small sample that rank was correlated with opportunity to copulate (table 5.4).

The main reasons for this lack of correlation between rank and

TABLE 5.4. Successful and attempted copulations between adult males and females of the Chippaberi troop during twenty-eight hours of observation between August 21 and September 5, 1971.

Male[a]	Successful copulations	Attempts to mount and copulate	Times harassed
Scarface	2	1	2
Big Plain[b]	2	1	2
Misshapen	3	0	3
Left-lip-scar	0	1	1
Jag	1	3+	3
Newcomer	1	0	1
Not identified	—	3	3
Total:	9	9+	15[c]

a. Ranked in order of ability to displace other males over food or position.

b. I suspected that Big Plain was the Chippaberi alpha male prior to the August 1971 invasion, but a definite identification was never made.

c. The bulk of this interference was enacted by Jag and Misshapen, though other males, females, and juveniles also participated in harassment.

opportunity were the willingness of females to solicit low- as well as high-ranking males, and the wiliness of low-ranking males who were frequently able to copulate, or at least to begin to copulate, out of sight of other animals who might hinder their efforts. (Even alpha males occasionally leave the troop to copulate with their consorts unmolested.) For example, on August 24, 1974, Jag, the youngest, smallest, and most subordinate—as well as the most persistent —of the intruders, was able to successfully copulate with a Chippaberi female by taking advantage of a moment when a bus passed through the rest-stop, and the whole troop and its invaders were fed by tourists. Only Jag and his consort remained on the hillside above, out of view of other animals, not attending the banquet.

An invader in the Hillside troop was able to profit from similar circumstances. During February and March of 1973, five new males (Righty Ear, Kali, No-No Man, Bluebeard, and Pequeno) were present in Hillside troop, in addition to the alpha male Mug, six females and two immatures. The following episode is reconstructed from fieldnotes for February 14, 1973, and from a super-8 film record:

> On the morning of February 14, 1973, between 10:40 and 12:20, as Hillside troop foraged in their core area near Eagle's Nest, Righty Ear made 14 attempts to copulate with Itch. On each attempt, the copulating couple was harassed by one or more other males, by Bluebeard (12 times), No-No Man (2 or more times), and Pequeno (1 time).
>
> Under most circumstances (access for food, position, or females), Righty is individually dominant to each of the males who are harassing him. When he approaches, Bluebeard, the most persistent harasser, is a composite of aggressive and subordinate attitudes: flashing fear grimaces, Bluebeard crouches low to the ground so as to be able to spring, swivel, and flee, and then inches close enough to threaten the copulating pair—especially Itch, who is never dominant to Bluebeard. In order to escape Bluebeard, Itch rushes out from underneath the mounted male. When Righty follows Itch onto the balustrade of a house, Bluebeard dangles from the roof above, swatting at the pair with a minimum of danger to himself.
>
> As long as Righty is simply in consort with Itch, no male intervenes. The cue for intervention is the moment that Righty puts his hand on Itch's rump and raises himself on his back legs to mount her. At one point, all the langurs except Righty and Itch had left an *Anogueissus* tree because a woodcutter had climbed into the tree and was noisily hacking at the branches. The couple in consort, busy grooming one another, are ignored by other troop members, until the moment that Righty takes advantage of this isolated opportunity to mount Itch. At that instant, Bluebeard, ignoring the woodcutter, rushes back up the

tree to harass Righty from the branch above his head. As soon as Bluebeard leaves, Righty mounts again. Again, Bluebeard rushes up the tree to intervene.

Later in the day, however, Righty achieves better results from a similar strategem under more opportune conditions: at 4:40, the invading males initiate a raid on a small vegetable garden, about one kilometer outside of the normal home range of Hillside troop.[8] At first, the Hillside females hang back, but in the distance they can see the males feeding on the vegetables. First Itch, then Oedipa, then Bilgay, follow. Sol watches from a distance, and at 5:00 p.m. joins them. Pawless never leaves the home range. At 5:20, an irate gardener discovers the langurs and begins to pelt them with stones. Mug along with Kali, Nono, Bluebeard, and Pequeno escape to trees to the northeast of the garden. The females, however, flee in the opposite direction, back toward their home range. Righty follows them to a grove of *Anogueissus* trees just within the Hillside boundary. At 5:30, Righty, without interference from any other male, copulates repeatedly with Itch.

When more than one male invades the troop, competition between them can be fierce. Once such invaders have attached themselves temporarily to a troop, they may also defend troop females against more newly arrived "invaders." Though in many cases we suspected that wounds were inflicted by conspecifics, the following incident, rewritten from the fieldnotes of D. B. Hrdy, describes one of the few instances where we actually witnessed one male inflict a wound on another in a fight over access to an estrous female.

At 3:00 p.m. on September 3, 1971, the Chippaberi troop is feeding in the vicinity of the large banyan fig, which is a focal point of their core area, when a large male (Newcomer) appears on the periphery of the troop. A female from the troop approaches and grooms the male. By 3:21, five females have gravitated toward him. These females seem familiar with Newcomer and are apparently at ease in his presence.

8. According to Mohnot (1971a), it is typical for a garden raid outside of the troop's normal range to be led by males. Within a troop's home range, however, it is frequently females who lead.

9. Our observations on the Chippaberi troop were intermittent. Consequently, we had little historical perspective on troop members and the males who passed in and out of their lives. We were especially curious about the casual and intimate welcome that Newcomer received from troop females. Possibly, Newcomer was an ousted alpha male, or else a returning "native son." Additional evidence suggesting that some, but not all, males in the vicinity of the Chippaberi troop were former

3:42, Left-lip-scar, one of the younger males who has temporarily invaded Chippaberi troop, moves toward Newcomer, then returns to the banyan. 3:48, Scarface, the most dominant of the invading males, follows two females who head toward Newcomer. Scarface, Left-lip-scar, and Big Plain (possibly the resident male of the Chippaberi troop) move one by one into the same tree that Newcomer has climbed up. Newcomer is in the uppermost branches, the three other males beneath him. The three males grunt, and Newcomer grinds his teeth; otherwise, there are no indications of hostility.

At 4:05, Scarface leaves the tree to threaten Misshapen who is attempting to mount one of two females (A and B) who are in estrus that day. Having broken up the copulation, Scarface returns to the base of Newcomer's tree.

At 4:20, Left-lip-scar attempts to copulate with female A but is harassed by Jag. Misshapen (the largest but third-ranking male) then chases Left-lip-scar away from the female. Big Plain and Jag, the youngest male, move close to female B.

At 4:40, Newcomer approaches the main part of the troop near the banyan, but he remains at least 17 meters from any of the other males. A Chippaberi juvenile runs toward him, squealing, playing his own game of derring-do, which Newcomer ignores. The other males avoid eye contact, and everyone except juveniles pretends not to notice Newcomer.

At 4:41, Big Plain mounts female B, but this attempted copulation is thwarted by Jag and Misshapen, who race toward him. Afterward, both harassers present to Big Plain.

At 5:45, female B moves away from the central banyan 46 meters down the main road toward where Newcomer was last seen, though at this time he is not in sight. Minutes later, Scarface follows her. Out of sight from the other males, Scarface copulates with female B until Left-lip-scar and Jag, who have also followed, see them. Under surveillance, Scarface dismounts and canters toward the two interlopers, grunting.

Back at the banyan at 6:30—in the absence of Scarface, Left-lip-scar and Jag—Misshapen mounts female A. He is harassed (ineffectually) by adult females and juveniles until Big Plain decisively intervenes, displacing Misshapen altogether from the vicinity of female A. Jag is midway between the two focal females, each with her own entourage of interested parties. He is apparently trying to decide whether or not to follow female B, who is disappearing from sight with Scarface and Left-lip-scar in her wake. Suddenly, all three

---

members was the contrast between some male band members who were brazen about taking food from humans (as all members of the Chippaberi troop learned to be) and others who were extremely shy. As far as we could determine, age was not a factor.

males, Jag, Left-lip-scar, and Scarface, come running back to the banyan. Female A is being mounted by Newcomer. Scarface charges, but instead of the usual perfunctory chase, the two males meet head-on. For just under 90 seconds, they grapple and bite at one another, then both withdraw and sit staring at one another. A chunk of flesh is hanging from the base of Newcomer's left thumb and he is bleeding profusely. Newcomer moves to the periphery of the troop and sits. He is still sitting there at 7:05 when observations terminate.

Previous observers of langurs have generally assumed that the "primary motive" of attacking male bands was "to possess adult females" (Mohnot, 1971a), though other explanations have sometimes been offered for invasions by neighboring alpha males. For example, Sugiyama has suggested that this sort of attack "seems to aim chiefly at the confusion of the other troop" (1965).

In the following analysis, it is assumed that most encounters in which one or several males (regardless of whether these are nomadic males or males from other troops) approach a bisexual troop, they do so to obtain information concerning the reproductive state of females in that troop, or to up-date their information regarding the strength of the females' male protection; to steal copulations; or to take over the troop entirely. If these assumptions are correct, one would expect that troops would attract invaders more frequently when estrous females were present in the troop, and/or when the troop was for some reason vulnerable.[10] Data on the presence of estrous females at the time of the seventy encounters observed at Abu tentatively support this assumption (table 5.5). Sixty-one percent of the time that a philandering male left his own troop to visit another, one or more females in the invaded troop displayed estrous behavior during that day—that is, at least one female solicited an adult male by presenting to him and shaking her head. Similarly, on 71 percent of the days that nomadic males visited troops, one or more females exhibited estrous signs. The percentage of time estrous females were present was greater in both types of male inva-

10. One would also expect that the frequency of *repeat* invasions would go up at times when the troop either appeared vulnerable or when estrous females were present. For example, if males invaded on day 1 when two females happened to be in estrus, one would expect that the likelihood of another visit on day 2 would be greater than the overall likelihood of an invasion. This certainly seemed to be true at both Dharwar (Sugiyama et al., 1965) as well as Abu, where male invasions became very frequent during the breeding season. Nevertheless, a controlled investigation in which data on female receptivity and male invasions are collected on a daily basis over a long period of time has never been undertaken.

TABLE 5.5. Proportion of time that estrous females were present during 70 agonistic encounters between individuals belonging to different groups, from data collected for six troops at Abu, 1971-1974.

Type of encounter	Number of observed encounters	Location	Percent of time at least one estrous female present	
			observed[a]	expected[b]
Between a male from one bisexual troop and members of another bisexual troop	18	61% in overlap	61	20
Between a singleton male or male band and members of a bisexual troop	21	76% in core areas	71	31
Between two bisexual troops (involving males, females, and immatures)	26	88% in overlap	77[c]	54[c]
Between females from different troops (adult males either not present or not involved)	5	60% in overlap	40[c]	43[c]

a. This column presents a minimum estimate, especially in the case of the last two types of encounters where estrous females in the less observed of the two troops might go unnoticed.

b. (.028) X (mean number of females); see text for explanation.

c. This frequency represents the combined total of estrous females in *two* troops.

sions than in the case of either intertroop or all-female encounters. On these latter occasions, estrous females were present on the order of 30 to 40 percent of the time.

These percentages suggest that the presence of sexually receptive females increases the likelihood of a foray by males from outside the troop. In order to test this hypothesis, some measure is needed of the likelihood per day of one or more females in a troop of a given size being in estrus. Unfortunately, the necessary information was not recorded for the same troop on consecutive days at Abu. However, short-term data on this topic were recorded for a single troop at a study site to the northwest of Abu, on the Gangetic plain (Dolhinow, 1972: table 5.3).

Using the data recorded by Phyllis Dolhinow at Kaukori, a crude estimate of the probability that an estrous female was present on any given day has been provided in the last column table 5.5. This percentage was estimated by multiplying the mean number of adult and cycling subadult females involved in each encounter by an estimate of the likelihood of at least one female being in estrus as extrapolated from information on the Kaukori troop. In this troop of 54 langurs, including 19 adult and 3 subadult females, at least one estrous female was present in 69 of 120 days. Thus there was a 0.58 chance that on any given day there would be at least one female in estrus in a troop containing 21 cycling females. The probability can be computed for a troop of X number of females by multiplying by the factor 0.58/21, assuming a linear probability distribution.

Obviously, the data available for the above calculation are far from ideal; the extrapolation incorporates several potentially inaccurate assumptions. The model presupposes that estrus probability can be proportionally extrapolated. In fact, though, female fertility at Kaukori and Abu may not be exactly comparable. A more realistic ratio (if it were available) would take into account such effects as birth spacing, synchrony between females, and seasonal breeding peaks. More importantly, it is very difficult in such a case to separate the cause and effect of what these males are looking for, and what they find, since estrous behavior may be induced among some females in the troop by the proximity of invaders. It is not absolutely certain, therefore, that the greater likelihood of invasions when estrous females were present was actually due to information obtained by males.

One glimmer of hope in this extrapolation is that the observed and expected estrous frequencies are very nearly the same (40% vs. 43%) for encounters involving only females—just as one would expect them to be if the hypothesis that the presence of estrous females increases the likelihood of male but not female intrusions were correct and if the extrapolation were valid. Nevertheless, because of the highly tentative nature of this analysis, any firm conclusions must await the collection of relevant data. Until then, the hypothesis that male invasions are correlated with the sexual receptivity of females remains unproven.

Assuming that each invader who approaches a troop is attempting to maximize his access to sexually receptive females, the alternative strategies—haunting, attack-retreat, temporary joining, and take-over—have differential costs and quite different payoffs in terms of the overall reproductive success of the animals involved. Included in the final tally of success are not only the number of successful inseminations but the physical costs and potential limitations on future possibilities incurred by a policy of attack versus safer alternatives such as threat behavior, haunting, and retreat. This analysis is based on Lack's principle that reproductive processes serve the reproductive interests of the individual (1954), and on a recent refinement of this principle by G. C. Williams to show how "expenditures on reproductive processes must be in functional harmony with one another, and with the costs, in relation to the long-range interest" (1966a:687). This type of analysis has been more specifically applied to aggression by Geist (1971) and by Maynard Smith and Price (1973). In the analysis that follows, the cost and benefit of each of four invasion strategies, insofar as these can be identified from information currently available for langurs, will be outlined.

Variables that may affect the final balance of different behavior patterns will differ for each individual and may include such factors as physical condition (size, strength, age, fighting experience) perceptual ability (including the individual's ability to accurately gauge his opponent's strength and willingness to fight), bluffing sophistication, and life stage (including the individual's potential and past reproductive successes). These same factors as they apply to his allies, his opponent, and his opponent's allies will also enter into the balance, along with chance factors such as a missing alpha male.

Haunting, a strategy that stops just short of actually invading the troop and entering into combat with resident males, entails the least risk of any of the four alternatives open to extratroop males. The most plausible explanation for haunting is that it represents an investigative effort by outside males to obtain inside information about both the reproductive state of females and the physical condition of their male defenders.[11] In those rare instances where the alpha male is either temporarily absent, missing, or dead, the haunting strategy has an immediate payoff. For example, the only copulations between haunting males and troop females observed at Abu occurred when subadult males entered the School troop at times when Harelip was either off chasing other males or foraging with another female apart from the rest of his troop. Haunting had an even greater payoff for the male band at Jodhpur which discovered that the leader of the B-26 troop had died (Mohnot, 1971b) and

11. In addition to this skulking about in the close proximity of troops, the orangutan-like ranging pattern of langur male bands has the advantage of traversing the home ranges of a number of troops and hence of bringing males into contact with many different mating opportunities (see Rodman, 1973b).

for neighboring troop leaders who discovered that the leader of the 2nd troop at Dharwar had been removed (Sugiyama, 1966). Within weeks, outside males had assessed the situation, taken control of the maleless troops, killed unweaned infants, and copulated with sexually receptive females. The benefits from investigation in these two exceptional cases were potentially the same as for a take-over, but without the risk of fighting a powerful alpha male.

Barring such lucky eventualities as a missing alpha male, the haunting strategy is one of low risk but also, under most circumstances, of low yield. At Abu there were no known cases in which haunting males were injured. On the other hand, they were rarely successful in stealing copulations.

The risk of injury is much higher for a male (or males) who invades a troop and engages resident males in combat before retreating. For example, the Chippaberi invader Newcomer was able to briefly enter the Chippaberi troop, mount an estrous female, and copulate with her. Instead of fleeing when harassed by Scarface, a male who had temporarily joined the troop, Newcomer met Scarface head-on and was wounded. Mohnot describes a similar case in which attacking members of a male band successfully copulated with females who left the besieged troop to solicit them. The males in this Jodhpur instance were more successful than Newcomer insofar as none of them was injured. Nevertheless, in neither case did any invader have sole access to an estrous female. On the day that Newcomer invaded the Chippaberi troop, his consort had already copulated with several other males, including Left-lip-scar and Misshapen. Even had she been ovulating (which is not known), there was a low likelihood that she would be inseminated by Newcomer. In the Jodhpur example, the females first solicited their new leader, and approached the invaders only after they had been ignored by him (Mohnot, 1971b:187). In both cases, the invaders retreated in the end, leaving the females in the control of a resident, or at least temporarily resident, male.

Several possibilities might account for an attacking male's subsequent retreat. The most obvious explanation is that the invader misjudged either the strength of his opponent or his determination to fight. Another possibility is that the attack was a gamble; whether or not Newcomer won or lost from being wounded in the example described above would have to depend on how valuable that single copulation was to him.

Undoubtedly, in many cases the attack-retreat strategy is simply a false start. But in some instances this stratagem is far from abortive. The cumulative effect of attacks day after day may lead to a more fruitful strategy such as temporary joining or take-over. For example, a band of males who attacked the 30th troop at Dharwar were able to severely wound the alpha male before retreating. On the day of their third attack one of the invaders was able to chase the weakened leader from the troop at relatively little cost to himself. If repeated attacks drain the strength of the troop's defenders, then the risks run by invaders in subsequent attacks are decreased. In addition, after an initial attack invaders are better informed both of the alpha male's strengths and of the intentions of their fellow invaders.

When circumstances permit it, one of the most advantageous strategies for an extratroop male is to temporarily join a troop. Invaders traveling with the Hillside troop had frequent opportunities to mount females, while males engaged in haunting or attack-retreat strategies only rarely did. Furthermore, considering the amount of time these males spent in the vicinity of the troop's resident males and the amount of aggressive harassment invaders engaged in among themselves, remarkably few were injured. Nevertheless, the low frequency of injuries may have been dependent on the situation. In the Chippaberi troop, males with fresh injuries were observed on three of ten days that outsiders were seen traveling with the troop.[12] Two of these injuries (a bite-wound in the thumb, the other a shoulder wound; color fig. 5) were relatively serious; the third, a facial wound, was not serious initially but subsequently became infected. In contrast, in 29 days that five invaders were observed traveling with the Hillside troop in 1973, there were no signs that these temporary joiners ever suffered any injury as a consequence of their presence in the troop.

The circumstances under which an alpha male will tolerate temporary joiners are not fully known, though one would expect this to occur only when the alpha male has more to lose by attempting to chase them out than he does by tolerating their presence—for example, when the sum strength of the invaders approaches his own. Even if the resident male is stronger than the invaders, he might allow them to remain in order to dissuade a third party presenting a more serious threat to his control of the troop from entering. The amount that the resident alpha male actually stands to lose because

---

12. Since it was not absolutely certain that the leader remained in the Chippaberi troop during this invasion, it is at least possible that a band of males took over the troop rather than invading it.

of the presence of these invaders will vary according to the probability that they will eventually push him out and that any of the invaders will actually inseminate members of his troop. It was likely that the Hillside troop females copulating with invaders were already pregnant.

From the invaders' point of view, several factors limit the advantages to be gained from temporarily joining a troop. Aside from the fact that females may not be fertile, interference during copulation attempts are common. In order to copulate, a langur male climbs onto the female with his weight distributed between his hindlegs, with which he clings to the female's back legs just below the knee joint, and his forelimbs, with which he clasps her rump. A copulating male can be easily dislodged from his precarious perch if another animal yanks at his facial hair or limbs or if the female runs out (or is chased out) from underneath him. As many as 86 percent of the copulatory attempts in the Hillside troop were harassed by other males; intromission was achieved less then half the time (fig. 5.6a-c).

The cue for harassment by another animal is any deliberate move showing the intention to mount, such as a male placing his hands on a female's rump. Consequently, one stratagem open to males would be to remain in consort with a female without trying to copulate with her for as long as they were visible to other animals. This principle might explain the commonly observed behavior of a male who displaces a competitor from the vicinity of a female but then makes no immediate attempt to copulate with her. In 1971, for example, I watched Scarface displace Big Plain beside an estrous female, but then ignore her frantic solicitations. So long as the two animals simply sit near one another or groom one another, they will not be harassed. Thus, while making certain that no other male copulates with the estrous female, Scarface himself could await an opportunity to be out of sight.

The potential cost to the invaders goes up dramatically as the likelihood that they might succeed in usurping the troop also increases. The reason is that up to that point, the alpha male would normally have much more to lose—his harem—if wounded, and this amount would exceed that which the invaders had to lose if they were temporarily debilitated. Hence, a resident alpha male would be hesitant to engage in an all-out fight. However, once an alpha male is seriously threatened by a take-over, on average he would suffer the same high cost for not fighting as for fighting, even if he was seriously wounded some portion of the time. Consequently, one would expect fierce engagements between the alpha male and the most determined invaders.

This predicted determination of the alpha male to fight raises a special set of problems for the invaders. Previously, the advantage of an attack was theirs: they were more willing to attack than the resident male would be to fight back. Furthermore, the alpha male might also hesitate to leave his troop to chase invaders away because other males might steal copulations (or even attack infants) in his absence. With the situation nearly equalized, that is, with both parties disposed to fight, and assuming that the alpha male is about as strong as any one of the invaders, the attackers lose their tactical advantage and must rely on numbers. However, since invaders do not benefit equally from a take-over, the recruiting of multiple invaders who would remain committed to the assault as the risk of participating increased might become difficult.

In almost all observed cases, only one invader remained in sole control of the troop after a take-over was accomplished. There is only one known exception in which two males remained in the troop. Unfortunately, observation of this troop terminated before the full length of their dual tenure could be established (Sugiyama, 1967). Given that only one male will remain, it would scarcely be advantageous for a young male, or a male in poor physical condition with little chance of remaining in the troop afterwards, to risk participation in an attack against an alpha male fighting to remain in possession of his troop and against females fighting to defend their offspring. Such differential prospects could explain the variability in male band composition that has been reported at Dharwar (Sugiyama et al., 1965), Jodhpur (Mohnot, 1971b), and Abu. On the eight days between August 21 and September 5, 1971, that the Chippaberi troop was observed during male invasions, five males were present on five days, four males on two days, and six males on one day. Five males who invaded the Hillside troop in 1973 were not traveling together a year later, when Righty Ear was spotted still in the vicinity of Hillside troop, but traveling in the company of two previously unknown males. By 1975 Righty Ear had managed to take over first the Hillside and then the Bazaar troops, but as far as I could determine the other males had left the vicinity.

At all three study sites where male invasions have been reported, males do not participate equally in attacks. For example, Mohnot describes a band containing twenty-two males which made repeated attacks on the B-26 troop at Jodhpur. However, in any one attack, as few as four of the twenty-two males might actually be

Fig. 5.6. (a) A Chippaberi female solicits a male . . . (b) who mounts and copulates. (c) The copulation is harassed by a parous female who approaches and . . . (d) threatens the copulating couple. (e) In order to frighten away his harasser, the male must dismount.

involved. On both this occasion and the attack on the 30th troop at Dharwar, the male who was subsequently to become the new alpha male initiated the attacks and was the most aggressive male band member in carrying them out (that is, YA male 1 at Jodhpur, and L male at Dharwar).

Assuming that the more invaders there are committed to the assault, the greater is the likelihood of take-over, how does an invader with a good chance of remaining as the new alpha persuade a male with a poor chance of remaining to join an attack that might have damaging consequences? There are at least two possible answers. In the case of Splitear and his eight young followers, there was a high probability that each of these young males was related to each other as half- or full-brothers, and to Splitear as sons. Hence, one would expect some cooperation between related members of male bands on the basis of kin selection. In terms of his inclusive fitness, a male might benefit by installing his brother as an alpha male. Another possibility is that the relevant information with which accomplices could assess their actual chance of remaining as alpha male is being withheld by the most powerful male.

The possibility that males are deceiving one another may explain a curious discrepancy between Sugiyama's observations of male bands and my own. Sugiyama and his coworkers have written that males in bands "form a group whose organization is not so highly developed" (Sugiyama et al., 1965). In contrast, I detected a stable hierarchy among members of a male band that had temporarily invaded Hillside troop (Blaffer Hrdy, 1974). Perhaps the discrepancy arises because Sugiyama was describing a band before its entrance into a troop, whereas my observations were made after the band of males had penetrated a troop. Given a situation in which success depends upon a joint effort but only one animal benefits in the end, it might be advantageous for the strongest animals to disguise their potential until after the band had entered the troop. This sort of deceit might explain why even so experienced an observer as S. M. Mohnot could be convinced that another langur besides YA-1 was the dominant male of the band. Nevertheless, it was YA-1 who initiated the attacks on the B-26 troop, who subsequently displaced all other members of his band, and who finally prevailed as the new alpha male of B-26. Mohnot writes: "The aged males and the dominant leader followed leisurely and were comparatively less excited [during the attack]" (1971b). This interpretation is obviously complicated by the endless possibilities for subtle twists; for example, once a male has "played his hand," how does he backtrack if he fails? In summary, manifestation of the male hierarchy may change with circumstances even though there is no basic change in the males themselves.

The Curious Case of Adulterous Solicitations

A mickle truth it is I tell
Hereafter thou'st lead Apes in Hell.
For she that will not when she may,
When she will, she shall have nay.
               Booke of Fortune, ca. 1560
               Cited by E. Kuhl, 1929

There are at least two ways to view copulations between females and extratroop males. From the point of view of the invaders, one might say that these are copulations "stolen" from the resident male; from the perspective of the female, one might see these as "adulterous solicitations." Though presumably males haunting the outskirts of a troop and males who actually invade the troop are awaiting opportunities to steal copulations, only one copulation was ever observed in which the female did not first quite clearly solicit the male—that is, present to him and shudder her head—and that exceptional copulation was by a resident alpha male who walked up to a seated female member of his harem, pushed her to her feet, and mounted her. The highest frequency of adulterous solicitations occurred when extratroop males had invaded the troop or were temporarily traveling with it (summarized in table 5.6). Outsiders, including members of male bands or, more commonly, the alpha male of a neighboring troop, were also solicited (table 5.7). At such times a female might wander some distance from her troop in order to solicit the outsider at closer range; if detected, she would be threatened, chased, and slapped at by the resident alpha male who would attempt to herd her back to the troop.

One explanation for these adulterous solicitations is that females are acting so as to pass on the genetic benefits of outbreeding to their offspring. This explanation is consistent with the finding that the highest incidence of extratroop adultery, in a small number of hours of observation, was reported for the School troop, which was politically the most stable of the troops studied at Abu and for this reason probably the most inbred. In most troops, given the pattern

TABLE 5.6. Attempted and successful copulations between soliciting females and males temporarily traveling with the troop.

Troop and date	Special circumstances	Invading males	Solicitations observed	Attempted mounts and successful copulations	Hours troop under observation that season
Chippaberi (August and September 1971)	Alien males enter troop for variable periods of time	Misshapen Scarface Left-lip-scar Jag Newcomer	Numerous	3[a] 3 1 4+ 1	77
Hillside (February and March 1973)	Five additional males have temporarily joined troop	Righty Ear Kali Bluebeard No-no Man Pequeno	Numerous, especially by Oedipa and Itch, presumed to be pregnant	35 8 7 0 2	156

a. See also table 5.4.

TABLE 5.7. Sexual solicitations of males outside the troop.

Troop and date	Special circumstances	Males solicited	Solicitations observed	Observed mounts	Hours troop observed that season
Hillside (August 20, 1971)	Shortly after Shifty usurps troop	Alpha male of I.P.S. troop	2 by Harrieta, who had recently lost an infant	0	42
School (December 29, 1973)	None apparent	Cast-eye, Blind-ear, and No. 3, all recently ousted from Toad Rock troop	Several by young adult female	2	24
School (January 1, 1974)	Harelip temporarily absent from troop	Cast-eye, Blind-ear, and No. 3	Several by young adult female	3[a]	—
School (April 21, 1975)	During encounter between School and Toad Rock troops	Alpha male of Toad Rock troop (Toad)	1 by young adult female	0	11
School (April 22, 1975)	None apparent	Alpha male of Bazaar troop (Mug)	Several by young adult female, as above	1[a]	—
School (May 10, 1975)	During encounter between School and Toad Rock troops	Alpha male of Toad Rock troop (Toad)	1	0	—

a. On both these occasions, squeals from the juveniles harassing the adulterous couple alerted Harelip to their activities.

of male take-overs about once every few years, a relatively constant influx of new genetic material is virtually assured.[13] But in the case of the School troop, Harelip had managed to remain in residence for at least five years (based on the composition of the troop when first encountered in 1971). This male with the defective lip probably fathered a young adult female and one, or possibly two, juvenile males who shared this defect, as well as other animals in the troop under five years of age. Therefore, solicitation of extratroop males by a young School troop female could reflect a bias among langurs against situations that would lead to close inbreeding. The fact that Harelip consistently interfered with attempts by young females suspected of being his daughters to copulate with males outside the troop could be explained if these females (whose progenitive opportunities might be fewer than Harelip's) had relatively more to lose from a pairing of deleterious genes in their offspring than Harelip did.

Nevertheless, there are several reasons why a quest for outbred vigor is not a sufficient explanation for all instances of adultery recorded at Mount Abu. Some gene flow was virtually assured both by the brief breeding tenure of alpha males and by the early departure of young males from their natal troop. Second, in several cases both at Jodhpur and at Abu, estrous females presented to outside males only after first being ignored by the resident male. Third,

13. Macaques provide an interesting parallel to langurs in this respect. In ten years of longitudinal data collected for one troop of rhesus macaques on Cayo Santiago island, four years—about the time it takes a female to mature—was the maximum that any one male remained (Sade, 1972). Japanese macaque males are even more transient (Sugiyama, 1976). In addition to the degree of outbreeding insured by the transience of resident males, at least 29 percent of offspring born in one troop of Japanese macaques whose blood proteins were examined must have been sired by males other than residents; these outsiders presumably entered the troop for a brief period during the mating season (Shotake and Nozawa, 1974, cited in Sugiyama, 1976).

Color fig. 1. The stable core of langur social organization is overlapping generations of female relatives accompanied by an adult male who enters the troop from outside it.

Color fig. 2. Somersaults and handstands have a fanciful appearance, but the enjoyment of such activities may be important in the development of motor skills. Elf (aged 3-4 months) watches as Quilt (13-14 months) does handstands. Quilt wears a play-face which appears to be a modified version of the langur fear-grimace.

1

2

3

4

Color fig. 3. Fights between long-time neighbors are predominantly bluff. Here Splitear and Harelip lunge at one another during an encounter in the overlap zone between their two troops.

Color fig. 4. Splitear escalates his bluff to a leap but avoids physical contact by landing far short of his opponent.

Color fig. 5. A young adult male (Jag) displays a shoulder wound he received while an invader in Chippaberi troop. The injury was probably inflicted by another male.

Color fig. 6. Despite the infant's resistance, it is forcibly pulled off one allomother onto another.

Color fig. 7. Itch did not desert her gravely wounded infant (morning of September 10, 1972).

females solicit outsiders even when they are not ovulating. In 57 of 69 adulterous copulations that were witnessed between troop females and outsiders, the troop leader had previously ignored solicitations by these females, suggesting that they may not have been maximally receptive to the alpha male at that time (though it may also have been important that in many of these cases more than one female was simultaneously in estrus). In at least 46 instances when Itch and Oedipa solicited invaders in the Hillside troop in 1971, they were almost certainly pregnant at the time. In other words, insemination may not be the object of such solicitations.

Related to the outbreeding hypothesis is the possibility that females are able to discriminate between male phenotypes and are selecting a "superior" one (see for example Ehrman, 1972, for drosophila). But this argument, like the previous one, is weakened by the fact that the females were first ignored by the alpha or were not ovulating in most cases. The possibility that females are responding to a limited capacity on the part of the alpha male to produce enough sperm to inseminate all fertile females in his troop must also be ruled out in cases where the adulterous female is not actually ovulating.

A further explanation as to why females solicit outsiders is that they are investing in the survival of future offspring by initiating consort relations with a variety of males, including the alpha male and potential invaders or usurpers; assuming that males remember the identity of previous consorts, they might be less likely to kill an infant that is possibly their own. To date, however, no satisfactory answer to the question of why females solicit outsiders has been presented in conjunction with data to conclusively prove it.

Summary of Territorial Options

As Suzanne Ripley has pointed out, if territoriality is defined either as the "defense of an area by fighting" or as the "exclusive ecological exploitation of some part of the home range maintained by a variety of territorial means," then some langurs are certainly territorial (1967).[14] At all three study sites where intertroop encounters are frequent (at Polonnaruwa, Dharwar, and Abu), the

14. In a recent paper, Vogel (1974) maintains that "a territorial basis for antagonistic intergroup encounters of Presbytis entellus is highly questionable." It should be noted, however, that Vogel's reassessment was based on only two observed encounters and that in each case these were attacks on a bisexual troop by male bands rather than intertroop encounters.

majority of encounters occurred in areas of overlap between troops or at the boundaries of exclusively used core areas; members of both sexes and all ages participated in this enterprise.

Assuming, however, that defense of the troop always entails some risk or cost to each individual defender, one would expect each langur to opt for participation only when some potentially greater penalty is attached to tolerating infringement of the troop's territory. The record of participation by males and females at Abu in conflicts involving members from different groups bears out this expectation. Males, females, and immatures all have a stake in the successful defense of troop boundaries or the successful invasion of territory occupied by another troop. Whether they are probing beyond normal home range borders for seasonally preferred foods or defending right of access to a potentially crucial foraging area, all troop members stand to gain if access to resources is won, or to lose if they are not.

Whereas troops defend their territories and almost never enter the exclusively held portions of their neighbors' ranges, such territorial defense and mutual respect of the territories of others does not apply to extratroop males. These nomads regularly trespass upon the territories of bisexual troops and on occasion even invade the troops. At such times, alpha males—but not necessarily other males or other troop members—may defend the troop's integrity or immediate space. So long as the invaders stay clear of the troop proper, the alpha male rarely leaves his troop to chase these itinerants out of his territory.

Haunting by nomadic males seems to be an investigative strategy; however, as the likelihood that these haunters might invade the troop and inseminate a female in the troop goes up, resident males become increasingly antagonistic towards them. Similarly, the alpha male's incentive to fight grows as the chance of a take-over by invading males increases. Confronted with the possibility of a take-over, resident females may also oppose the invaders, since incoming males unrelated to troop infants may kill them.

Extratroop males have less to lose in a fight than do males already residing in a troop, and hence are more likely to attack a troop than vice versa. Within a band of extratroop males, there is great individual variation, and the strategy of each varies accordingly. After a take-over, only one male will remain as the new alpha male; one would expect males with a good chance of prevailing to have the most sustained motivation for attacking bisexual troops—especially as the risks entailed in fighting a determined alpha male go up.

All troop members share the area containing physical resources used by the troop. Resident males and most females share an interest in defending infants born in the troop. Only males who have a good chance of inseminating females in the troop, however, have much at issue in defending reproductive access to troop females. In most cases, only one animal, the resident alpha male, stands to lose much from copulations stolen from his harem. Within these three categories—the larder, nursery, and harem—assessment of the price of participation in a conflict with members of another group will vary from case to case. For an adult female, reproductive state, such as whether the female is pregnant, carrying a newborn, accompanied by an unweaned infant, cycling, or in estrus, may be paramount. Other factors include her degree of relatedness to other threatened troop members. Males must take into account the strength of a male opponent, the likelihood of injury, and, in the case of an alpha male, the actual probability that an uncontested invader will inseminate a member of his harem. Far from being a unitary response, "territoriality" and troop defense reflect a wide range of situation-dependent motivations concerning each individual's willingness to attack another troop and defend his or her own.

6 / Competition among Females

On an average, each Indian car manufacturer receives as many as 10,000 complaints every year from dealers and owners. Enquiries reveal that an Indian car lies in the workshop for about 30 to 45 days a year . . . What are the causes of the steady deterioration in the quality of Indian cars? Enquiries reveal interesting facts. Car manufacturers use substandard ancillaries, raw materials, finished components and other equipment. The car industry depends on outside sources for 40 to 50 per cent of its ancillary requirements . . . Such being the case, relatives and close associates of car manufacturers are encouraged to set up automobile ancillary units . . . The dealers enter into a special arrangement with manufacturers for buying ancillary items from those units which turn out cheap and substandard components.

. . . Yet, the manufacturers pay exorbitant prices for shoddy spare parts as they are interested more in the prosperity of the ancillary units owned by their kith and kin than in making good profits for themselves and safeguarding the interests of their shareholders.

From "Our Rattling Good Cars,"
Times Weekly, vol. 4, no. 25, supplement to
The Times of India, January 20, 1974.

William Hamilton's theory of inclusive fitness predicts that degree of relatedness should have a crucial bearing on the likelihood that any two individuals will behave altruistically or selfishly toward one another in any given situation (1971). If, for example, individuals in the same troop are on average closely related, we would expect rewards of intratroop aggression to be devalued by the risk of injuring near relations. If there is not comparable risk in intertroop encounters, we would expect fiercer aggression, even though exactly the same resource might be at stake. By the same token, we would expect the largest amount of cooperative and altruistic behavior to take place among individuals in the same troop. Cooperation, defined as coordinated behavior by two or more individuals regardless of whether both benefit in the same way from the common achievement, may be either partially or entirely self-serving, at least in a genetic sense, though it need not be so. By contrast, competitive behavior, or the attempt by one individual to monopolize prerequisites of survival and reproduction, is by definition selfish since personal fitness is increased at the expense of another individual's fitness. Selfish behavior consistently directed toward a weaker individual who is unable to retaliate is labeled exploitation.

Insiders and Outsiders

Among langurs, competition is most striking in the contest between males for reproductive access to females. Nevertheless, competitiveness is also discernible in relations between females from different troops, in the typically mild conflict that occurs between females belonging to the same troop, and in the existence of a ranking system among female troop members. By and large, the repertoire of aggressive behaviors is the same whether the hostility is between females of different troops or between members of one troop, but there are differences in which elements of the repertoire are emphasized. The basic female inventory of aggressive behaviors, arranged more or less in an ascending order of intensity, reads: *approach*; *threats*, including staring at the opponent and the gamut of facial expressions ranging from grimaces and open-mouthed threats to biting air accompanied by head-bobbing; *attack*, ranging from mild imitation of true attacks, such as lunging in place or slapping the ground while remaining in place facing the opponent, to true attacks, which include charging and slapping, hair-pulling, pinching, shoving, tail-biting, forward lunges that occasionally entail grappling or biting, four-footed jumps onto the back of an opponent, either while the opponent is stationary or during a chase; or some combination of the above.

The most frequently heard female vocalizations during fights (as well as at other times) include grunts (an "agh" sound), short barks, and squeals. Two other sounds are made only during fights or at times of generalized tension and excitement: teeth-chattering (which may be the more modest, female version of the rasping canine-grinding of males)[1] and a nonresonant version of the full male whoop, a whistling "whrr-whoop" sound, possibly equivalent to what Jay (1965) calls the "female whoop." The typical pose for females during an intertroop encounter is facing the opposing troop, lunging, and slapping at them (fig. 6.1). When several females from the same troop participate, it often appears that the females have lined up for some ritualized form of team fight.

Although intertroop encounters involving whole troops are initiated by alpha males at least half the time, occasionally (in five of twenty-six encounters) it is a female who is the first to fight with members of the opposing troop. In addition to the twenty-six encounters between troops, five encounters were initiated by females and involved only, or primarily, females. On one of these occasions, no adult male was present in either troop (since one male, Shifty, was off chasing the other, Mug). In three other incidents, one male was present in one of the two troops but took little part in hostilities. In the fifth incident, involving the intertroop kidnapping and attempted retrieval of an infant, males were present in both groups, but were apparently unaware of the frantic maneuvers by females in their harems.

These primarily female engagements were unusual in that four of five occurred between females in Hillside and Bazaar troops at times when one male was attempting to retain control over both troops at once. Despite the fact that Hillside and Bazaar troops sometimes shared the same alpha male, each troop continued to live in its distinct home range with only a limited area of overlap between them. In both cases, the double-usurper was spending most of his time with the larger Bazaar troop in its home range, making only sporadic visits to Hillside troop. In 1972, however, Shifty's bid for control over both troops was complicated by a rival male, Mug, who would enter Hillside troop and stay there in Shifty's absence. In 1973 Mug was joined by five additional males; all six fled, however, whenever Shifty approached Hillside troop. The following sum-

---

1. As far as I know, this teeth-grinding sound has never been reported for langur females living in the wild. However, Ripley mentions that "captive juvenile females in the experimental colony of the Hooper Foundation, University of California Medical Center, commonly gave vigorous tooth grinds as part of their generally abnormal behavior patterns deriving probably from the very restrictive caging and social deprivation" (1965).

Fig. 6.1. Females in an intertroop encounter line up facing the opposing troop, lunging and slapping at them. Here the two adult males (LeGrand standing, Shifty seated) remain on the sidelines while the females engage in their slapping ritual.

marizes encounters between females in Hillside and Bazaar troops during this politically complex time:

> At nine in the morning on September 2, 1972, the Hillside troop dozes and grooms in the same tall *Grevillea* grove in which they spent the night. Nearby, Bazaar troop feeds in the same large fig tree in which they spent the night. Both sleeping sites are situated in the overlap zone between the home ranges of the two troops. Shifty Leftless (with Bazaar troop) is the only adult male present.
>
> At 9:30, 9:50, and again later in the day at 1:58 p.m., two to three Bazaar troop females make forays into the *Grevilleas* occupied by the Hillside females. On each occasion, the invaders are chased back by Hillside females. In the course of these raids, the most active offensive and defensive roles are played by older females: Short from Bazaar troop, Pawless and Sol from Hillside troop.
>
> At 10:25 on the morning of September 8, 1972, Mug and the Hillside females leave their sleeping trees in a *Grevillea* grove well within Hillside troop's exclusive home range and move towards "Boulders," an estate in the overlap zone used by both Hillside and Bazaar troops. The Bazaar troop is foraging to the west of Boulders, out of sight but not out of occasional hearing, and is gradually moving towards Hillside troop.

On the day preceding this encounter, Mug had made at least three attacks on Hillside mothers with infants. On the day following this encounter, one of the Hillside infants was attacked and severely wounded by Mug. At 11:00 on this day, Mug contacts a Hillside mother and is seen momentarily grappling with her before withdrawing. Though I did not witness the complete sequence of events, it is probable that Mug had threatened her infant. (Such assaults on infants may explain why Hillside females seemed to be lingering in the overlap zone, as close as they could be to Shifty and his new troop without leaving their own range.)

At 11:20, Mug precipitously leaves Hillside troop just moments before Shifty enters it. Accustomed to Shifty's visits, the females are calm and unprotective of their infants in his presence. Shifty ignores Hillside females, climbing onto a granite outcrop to stare intensely to the northeast in the direction that Mug has disappeared. Seconds later, Shifty runs towards the thicket in back of Hillside House in pursuit of Mug. Ten minutes later, Shifty reappears in Hillside troop, which is foraging in the overlap zone. Shifty forages with them for a period, then returns to Bazaar troop.

Some two hours after Shifty's departure, at 1:42 p.m., a middle-aged Bazaar troop female (Kasturbia) suddenly approaches the Hillside females in the overlap zone. Kasturbia chases Bilgay, almost, but never quite, catching her (fig. 6.2). The only sound noted was a distinctive teeth-chattering. At 1:45, Kasturbia contacts Itch by jumping four-footed onto the Hillside mother's back. Kasturbia then chases three other Hillside mothers with their infants clinging to them across a field. Just as suddenly as this chase began, Kasturbia veers off in pursuit of a fifth Hillside female (Sol), chasing her eastward into Hillside troop's core area. As Sol disappears from sight, Kasturbia turns on the sixth Hillside female, Pawless, and chases her up a hill into thick lantana bushes. As the two females lose sight of one another in the undergrowth, Pawless rears up on her hindlegs to look around; Kasturbia climbs into the low crotch of a tree.

Kasturbia descends and returns to the field where Bilgay has been watching this chase. Instead of running away, Bilgay presents to Kasturbia. The trespassing female then inspects Bilgay's five-month-old infant. Bilgay responds by clutching her infant tightly and turning her back to Kasturbia. Again Kasturbia reaches for the infant. Bilgay moves in back of Kasturbia, and for a few seconds grooms her. Kasturbia momentarily leaves Bilgay to approach Harrieta, who grunts, · but who does not run away. Kasturbia then returns to Bilgay and is groomed by her for five minutes. At 2:21, the invader from Bazaar troop looks off to the west in the direction that her own troop should be, and heads that way.

Fig. 6.2. The Bazaar troop female Kasturbia unrelentingly pursues Bilgay and her infant deep into Hillside troop's core area.

In summary, this single Bazaar troop female spent 43 minutes with the Hillside troop in what was previously thought to be their exclusive home range, harassing and chasing them; shortly before leaving, the trespasser approached the mother of the youngest Hillside infant and was groomed by her. The trespassing female made several unsuccessful attempts to hold the infant.

Five months later, antagonism between females in the two troops persisted:

> Just before noon on February 15, 1973, Mug and the five males temporarily traveling with Hillside troop are chased out by Shifty. None of these males rejoins either troop before evening. In the absence of any adult males, females in Hillside and Bazaar troops drift together and converge at about 1:00 p.m. in the yard of Jodhpur House. Youngsters from the two troops play and play-fight. Around 1:30 some children throw stones at the langurs and a Hillside female (Pawless) charges them. Moments later, females in the two troops begin to chase one another. The two troops form a more or less single line facing one another, break apart, and then confront one another again. Contacts on this occasion include slapping, grappling, tail-biting, and one female jumping on top of another.
>
> At midday on March 6, 1973, Mug and the five who have temporarily joined him in Hillside troop doze in a mango tree in the vicinity of

the Hillside females in an area normally used exclusively by the Hillside troop. At 1:50, the males start awake; there is unisonous urinating, defecating, grunting, and teeth grinding by them, accompanied by powerful fecal odors. Five minutes later one of the males whoops and displays by leaping rambunctiously about, rattling the branches of the mango. On a wall not one hundred meters away, Shifty appears grimacing, grinding his teeth, biting mouthfuls of air. The six males rapidly leave the vicinity, and are not seen near Hillside troop again that day.

By 3:00 Shifty has rejoined the Bazaar troop near the Bazaar. Gradually, the troop drifts eastward, and at 5:45 they converge near the Boulders, more or less equidistant from where both troops started out. Hostilities begin when several Bazaar troop females chase Itch up a tree. Shifty lunges at Pawless who slaps back at him. After this, Shifty ignores the females' activities. He climbs into a tree, grinds his teeth, and moments later whoops and displays.

Shifty descends. Harrieta's fifteen-month-old son approaches within four meters of him, making mock charges. An estrous subadult from Bazaar troop (Guaca) then chases the juvenile back to his mother. As Pawless and two Bazaar troop females come up to support him, the juvenile male attacks Guaca and this handicapped young Bazaar troop female retreats before him, presenting and shuddering her head.

The circumstances surrounding these encounters, and the initiative taken by Bazaar troop females in instigating hostilities, suggest that Shifty's attentiveness to Hillside troop during 1972 and 1973 may have provoked antagonism toward females in Hillside troop from females in Shifty's new troop.[2] In three of the four encounters, Bazaar troop females attacked Hillside females within hours after Shifty had visited then and chased out Mug. The antagonism between females in these two troops is of particular interest since it would have been both to Shifty's advantage and to the advantage of Hillside troop mothers for the two troops to merge. As it happened, however, the hostility of Bazaar troop females made such a merger difficult, if not impossible. Failing a merger in either 1972 or 1973, four infants in the highly vulnerable Hillside troop were attacked by Mug and other male invaders. Since Shifty was

2. Mohnot has reported three similar encounters at Jodhpur that occurred when females without an alpha male (the decimated B-26 troop) met a neighboring troop (B-25). In each case only adult females, or adult females, subadults, and juveniles (including two juvenile males), entered the B-26 troop while the male leader and females with young infants watched from a distance (1971b).

probably the father of some, if not all, of these infants. Shifty as well as their mothers suffered reproductive losses from Shifty's inability to keep Hillside females under protective custody.

As in chapter 5, only encounters from the first four study periods were analyzed. It is interesting to note that of two additional all-female encounters in 1975, one of these was also between females in Hillside and Bazaar troops at a time when a double-usurper (this time Righty Ear) was trying to retain control of both. To a much greater extent than Shifty had, Righty Ear took an active role in promoting a merger between the two troops:

> On the morning of May 19, 1975, Bazaar troop and a maleless Hillside troop meet near Jodhpur House in the area of overlap between their ranges. As a Hillside female who is in the last weeks of pregnancy (Bilgay) approaches, she is vigorously attacked by a young Bazaar troop mother (Guaca) who jumps three-footed onto Bilgay's back and then chases her. Bilgay flees, running up to Righty Ear and sitting beside him. As Bilgay moves away from Righty moments later, Guaca approaches her, and Bilgay again sits beside Righty, who has been watching the spatting females and grunting.
>
> At 9:01 Righty pulls Bilgay toward him. For three minutes, he grooms her. Then for nine minutes, she grooms him, followed by Righty grooming her for four more. During this period, Righty is extremely attentive to Bilgay, to a degree which is almost never seen among langurs except when a male is consorting with an estrous female.
>
> Shortly after, pregnant Bilgay moves away from Righty; she approaches a Bazaar troop female, emitting short squeals. Within moments (at 9:50), several Bazaar troop females, including Guaca, bear down upon the Hillside troop intruder. Guaca leaps onto Bilgay's back, flailing at her momentarily. Squealing, Bilgay retreats towards Righty; he is grunting and biting air in the direction of the spatting females. Righty charges Guaca, then returns to Bilgay. Once again Bilgay and Righty sit side by side. Two minutes later Righty leaves Bilgay to make another charge on the Bazaar troop females who have been harassing her.
>
> At 10:10, 11:00, and 1:05, Bazaar troop females make attacks on Hillside females followed by counterattacks from Righty. In the course of these disputes (but not actually witnessed by us) Bilgay's upper lip is split open; a narrow strip of skin and flesh dangle down over her mouth.

During the succeeding weeks, my observations of Bazaar troop were curtailed by surveys elsewhere. Nevertheless, there seemed to have been few substantial changes in relations between the two troops.

> In the afternoon of June 19, 1975, Bazaar and Hillside troops meet and mingle in Jodhpur yard. Since she was last seen, Bilgay has given birth, and somehow received a wound on her side. Shortly after the two troops come together, Elfin and Guaca chase Hillside females, slapping fiercely and lunging at them. These events are followed two hours later by Righty assaulting females in the Bazaar troop. Since his attacks on Bazaar troop females are so frequent at this time, there is no way to know if his behavior is related to their earlier treatment of Hillside females.

Despite what I interpreted as a definite policy on Righty's part to offer safe shelter to Hillside females, a merger between the two troops never occurred. When we departed from Abu on June 20, Righty's two harems were separate entities. Except for occasional, invariably antagonistic, meetings between the troops and for brief visits by Righty to Hillside troop, the remaining Hillside females and their offspring wandered maleless in their own home range.

The final all-female encounter to be described has nothing to do with males. It occurred when a subadult female (Junebug) from the Bazaar troop kidnapped a newborn infant from the School troop. The following is largely derived from a filmed record of the events made by D. B. Hrdy:

> Beginning at midday on December 23, 1974, and continuing over the next three hours, the Toad Rock and School troops confront one another just within the border of the School troop's core area. Periods of chasing and fighting alternate with periods of relative calm. During this encounter, a School troop mother with a newborn infant (one-to-two-days-old) is seated on a haystack well to the east of any fighting, in a corner where the ranges of School, Toad Rock, and Bazaar troops all converge. This particular mother is young, possibly primiparous, and very restrictive of her infant. Earlier on this same day, the mother has rebuffed five of seven attempts by other School troop members to take her infant.
>
> At 4:40 the young mother is approached by a subadult female (Junebug) from Bazaar troop. The two females at first avoid eye contact. Then the mother threatens Junebug. Momentarily both females slap at one another. Junebug reaches for the infant, catches hold. There is a violent tug-of-war in which the two females tumble together down the haystack, each refusing to let go of the infant. Junebug apparently bites the infant. The mother releases her grip and

Junebug runs off bipedally, holding the infant to her body with one hand. As the School troop mother pursues Junebug, she is intercepted by an old female from Bazaar troop (Short). Short is joined by a second old female, Quebrado. Together they chase the mother back toward School troop. At a roadway, the mother hesitates, making several false starts towards the direction of the retreating Bazaar troop females. She disappears in the direction of School troop's core area.

Meanwhile, the abducted infant has been carried by Bazaar troop members to their sleeping site near Sanand House. The infant is passed around among various Bazaar troop females, including a 13-month-old cream infant. By the following morning, the mother is once again in possession of her infant, but the actual retrieval of the infant from its captors was not witnessed.

Similar kidnappings by both adult and subadult females have been reported at Dharwar by Kawamura and Yoshiba (cited in Sugiyama, 1965a) and by Sugiyama (1966), and at Jodhpur by Mohnot (1974). In the incident observed by Kawamura and Yoshiba, a subadult female from Dharwar's 1st troop kidnapped a one-day-old infant from the 7th troop. In this instance, both the mother and the "leader" male attempted unsuccessfully to retrieve the infant from the "overwhelmingly stronger" 1st troop. In the Abu case, only the mother tried to retrieve her infant, probably because all members of School troop except the mother were engaged in a confrontation with Splitear's band at the time of the kidnapping and were unaware of the mother's predicament.

In these two cases of infant-stealing—the School troop incident and that in the 7th troop at Dharwar—kidnapping had a divisive effect, sparking chases and other hostilities between females in the troops involved. In another case, however, repeated kidnappings from one another by females in Dharwar's 2nd and 3rd troops at a time when the alpha male of 2nd troop had been experimentally removed apparently promoted a merger between the two troops. As Sugiyama described it: "After the 2nd and the 3rd troop had a single leader in common, if infant-transferring took place between the troops, naturally the mother of the 3rd troop would remain in the other troop, following the tender of her infant, for there was no other offensive adult male in the 2nd troop except the leader of the 3rd troop . . . It is certain that the 2nd and 3rd troops were combined into one through the intertroop transferring of infants" (1966).

If Sugiyama's interpretation is correct, the powerful attraction exercised on females by newborn infants was able to bring about a merger between females in different troops under circumstances very similar to those which occurred at Abu in 1972 and in 1975 when (from one point of view) it can be said that Shifty and Righty failed.

In-House Squabbles

While chasing another female or charging at her, lunging, and slapping are typical of disputes between females belonging to different troops, within troop aggression is usually toned down to facial threats, ground-slapping, staring, pinching, and hair-pulling, acts that take place at close quarters and last just seconds (fig. 6.3). More rarely, a female may pick up a troop member's tail and bite it, or jump four-footed onto her back (fig. 6.4). Rewards to be gained from competing with a fellow troop member are devalued, but they are not erased. Family spats occur mainly in two situations: when one female displaces another, and during sexual harassment.

Sexual harassment has been reported for nearly every site where langurs have been watched, and it is obvious from figure 6.5 that this peculiar behavior has not escaped the attention of local inhabitants. As soon as a male mounts a female, other troop members come running over to slap at the pair or to attempt to pull the male from his perch or else to chase the female out from underneath him. While Jay reports that at Kaukori only "less dominant" males (1965) harassed the copulating couple, Ripley states that at Polonnaruwa both subadult males and females were involved (1965). At Dharwar, adult and subadult females and, in a few cases, less dominant males tried to break up copulations (Yoshiba, 1967). At Abu, participation

Fig. 6.3. Within the troop, close-quarters acts of "bitchiness" such as hair-pulling, pinching, or slapping are more typical than chasing.

Fig. 6.4. Occasionally, one female jumps with all four feet onto the back of a subordinate. Here, a young adult female grazes Pawlet's back as she performs an incompleted four-footed mount. Note Pawlet's deformed right paw.

Fig. 6.5. Carving on a shrine, Bhaktapur, Nepal, which depicts juvenile harassment of copulating monkeys.

was general. Except for the very young, males and females of all ages (juvenile, subadult, and adult) attempted to interfere at one time or another. The only sex difference that was apparent was that juvenile males were more likely to harass a copulation than females were, though in one exceptional case a juvenile female offspring (Mole's) was the fiercest participant in harassing her mother's copulations with Splitear in 1973. Individual involvement varied considerably with the composition of the troop. In 1973, almost all harassment in Hillside troop was by adult males harassing other adult males; during the same season, most harassment in Toad Rock troop was by juvenile males, though juvenile and adult females also participated.

Interestingly, relative status did not seem to affect the likelihood that an animal would attempt to interrupt another's copulation. Low-ranking females would attack the alpha female; nearly rank-less juveniles would assault an alpha male. Relative status did, however, make a difference in the form that harassment took. A less dominant male harassing a dominant animal would assume a crouching posture that could be quickly converted into flight while an infant harassing its mother's consort would do so from a position of relative safety. An infant-2 in the Chippaberi troop dangled from a vine above his mother's head, simultaneously squealing with fright and beating against the alpha male who had mounted his mother (fig. 6.6).

The only published paper to deal specifically with sexual harassment in stump-tailed macaques and other species (Gouzoules, 1974) offers the seemingly Freudian interpretation that harassment is an altruistic attempt by fellow troop members to deflect the male's "aggression" (which otherwise would be vented on his consort) onto themselves. There is no reason to assume that this is so, or even that the motivations for harassment by different ages and sexes are the same in each case. Adult male harassers are probably attempting to prevent a male competitor from inseminating a troop female. The motivation of immatures is less clear, but at least in the case of the female's own offspring, their mother's investment in them might be diminished by the birth of a sibling, a potential rival for maternal nurture. Harassment by other adult females is the most puzzling, however. In some instances, estrous females were the most vigorous harassers, but this was not necessarily the case. No evidence exists as to whether the insemination of one female reduces the likelihood of a second female in the same troop, either concurrently or subsequently in estrus, also being inseminated, but it is possible that in some cases male sperm may be a limiting resource. Another possi-

Fig. 6.6. An infant-2 swings on a vine above his mother and her consort, harassing the disgruntled male, who appears to resent the interference.

bility is that females are attempting to reduce future competition for resources by limiting births within the troop—an explanation which presupposes that there *is* competition among females for resources.[3] Assuming that this is the case, how would the female portion of the troop's resources be allocated?

3. Among African wild dogs, for example, where there is communal rearing of young, mothers may obtain a greater proportion of their pack's resources for their own cubs, including the milk of other females, by murdering the offspring of their competitors (van Lawick, 1973).

The Female Displacement Hierarchy

Hanuman langurs are not obvious candidates for a study of female dominance relations. At all other study sites for which published reports on dominance interactions exist (Kaukori, Dharwar,
and Polonnaruwa), the authors specifically state that females have
poorly defined and nonpredictable dominance relations. Jay, for example, found it impossible to assign a female langur to a position
within a dominance hierarchy except within a general level of dominance that includes females of approximate rank. Several authors
report that a female's dominance position fluctuates with the
different stages of her reproductive cycle (Jay, 1965; Ripley, 1965).
Both Sugiyama (1967) and Yoshiba (1968) report that dominance
relations at Dharwar were even less clearly defined that those at
Jay's site. Marler reports that among African black-and-white colobus females "no signs of a dominance hierarchy could be detected"
(1972). In contrast to all of the above, but in agreement with the observations at Abu, Poirier reports that in the Nilgiri langur groups he
studied in South India, the female dominance order was relatively
stable: "Even though encounters were rare, adult females could be
ranked in a linear order where relationships between females
seemed rather clearly defined. Certain females constantly dominate
others by assuming priority at desired feeding and resting locations" (1970).

In the course of the study at Abu, I witnessed many hundred displacements—one female approaching another, directing a threat or
contacting her (a cuff, pinch, hair-pull, and so on), followed by the
other animal moving away or relinquishing some position or
resource. Almost all of these displacements occurred in the following contexts: in competition over finite food commodities that
may be either naturally occurring or artificially provisioned by
humans; over feeding positions; over positions not directly involving
food (in which case it is difficult to determine what is actually at issue); over possession of an infant (usually displacement occurs just
after one female has tried unsuccessfully to take an infant from
another); or as part of the defense of a third animal—a rare event.
One exception to the general pattern of displacements is that a
female who firmly possesses a specific resource (for example, a
euphorbia stalk or an infant) may wrap herself around a resource
and refuse to give it up even though she is threatened by an animal
normally able to displace her (fig. 6.7).

A great number of the displacements recorded in this study oc-

Fig. 6.7. Mopsa huddles over her just-born infant (note umbilical cord by Mopsa's right rear foot) to avoid giving up her infant to the troop's pregnant alpha female, Mole.

curred over food provisioned by humans, either by the observer or by local people unrelated to the study. Naturally occurring food items that langurs compete for include mangos, especially the germinated sprouts from remnant mangos, and such items as ripe figs, which are available at times when most fruit is still green. Langurs also displace one another over positions that allow access to localized resources, such as a clover, sap-lick, euphorbia, or lantana berry feeding positions (fig. 6.8). Table 6.1 indicates those displacements for which the apparent incentive could be determined.

Undoubtedly, frequent feeding of langurs in this population is responsible for the high frequency of displacements and probably accounts for the generally high level of aggressiveness (Singh, 1969, for rhesus macaques). In recent years Thelma Rowell (1972b; 1974) and others have questioned the validity of data collected under conditions in which animals are provisioned, contending that ranking systems not present under "normal" conditions might be induced by artificial feeding. However, in Hindu-inhabited areas, provisioning is a normal part of langur and macaque existance and has been for at least several thousand years. It should also be noted that seasons

Fig. 6.8. At Abu, palm sap is used by local people to prepare a partially fermented drink called "neera." The sap is obtained by hacking through the bark at the top of the tree. For several days after tapping, sticky sweet sap rains from the treetop, coating the trunk and the tangle of rootlets at the base of the tree. Here members of the Toad Rock troop harvest the remaindered sap. Turn-over between feeding positions is brisk as dominant animals glean the more easily removed gum and then move to a new location, displacing whichever animal happens to be there.

TABLE 6.1. Contexts in which females from three troops displaced one another during 671 interactions recorded between June 1972 and June 1975.

Displacements	Bazaar troop		Toad Rock troop		Hillside troop	
	1973	1973-74	1973-74	1975	1972	1973
Artificially provoked (69% of total)	77 (71%)	54 (76%)	79 (52%)	130 (69%)	74 (79%)	48 (83%)
Naturally occurring (31% of total)	31 (29%)	17 (24%)	73 (48%)	58 (31%)	20 (21%)	10 (17%)
Over food	16	6	36	36	5	9
Over position and other	8	7	11	22	15	1
Over access to newborns	7	4	26	a	a	a
Total:	108	71	152	188	94	58

a. No newborn infants

of scarcity and competition for finite food commodities are built into langur life at Abu, and when confronted with decreased food sources as in the 1975 drought, langurs appear to resort to increased competition. Finally, the ease with which almost all social primates adapt to a competitive system when conditions make it advantageous to do so suggests that this response has been selected for in the past and is an innate part of their adapting repertoire.

Based on displacements drawn from the daily records for those events in which both partners could be recognized, females were ranked in order of their ability to displace other females. Roughly 30 percent of these displacements occurred naturally, while the remaining displacements were provoked by human provisioning. Separate matrices were constructed for each of the three troops for each two-to-three-month study period for which sufficient data were collected. Two rules were used to rank females along the axes of the matrices. Top priority was given to the "rule of least reversals." That is, females were arranged along the vertical axis of each matrix so that the fewest possible number of females were ever displaced by a female listed below them. Because some females never displaced each other, it was impossible using the first rule to give them anything but an intuitive assignment. To avoid this, a second rule was applied. The number of cells in which a female was ever displaced was subtracted from the number of cells in which she displaced another female. Females were then ranked with the higher position assigned to the highest score. Despite this second rule, some ties persisted. In view of the difficulty of ranking females who participated in few displacements, and possibly in line with a real-life situation, the relative rank of females lower down in the hierarchy might best be viewed as somewhat blurred.

Matrices for the Toad Rock troop arranged for the 1973-74 and 1975 study periods (tables 6.2 and 6.3) illustrate typical features of langur hierarchies. In general, high-ranking animals are over-represented in casual or ad libitum samples of displacements. This same phenomenon also appears to be true for langur male hierarchies (table 8.1) and for various other primate species (for example, rhesus macaques and savannah baboons). The alpha animal is often responsible for the greater part of displacements; the closer a female is to the alpha position without actually occupying it, the more likely she is to be displaced by the alpha female. Relations between females at the bottom of the hierarchy, on the other hand, were obscured by the reluctance of these females to compete in the presence of higher ranking animals.

Though the observed frequency with which individual animals participate in displacement activity may be affected by biases inherent in casual (ad libitum) sampling, the direction of displacements, if it is consistent, would not be (Altmann, 1974). From these matrices, it appears that the ability of one female to displace another is predictable from month to month. For example, in six matrices constructed for the three troops out of 578 female-female displacements in which both partners were recognized, females were displaced by females ranked below them on only fifteen occasions. In other words, the proportion of the data not amenable to the rule of least reversals was small, about 2.6 percent for the reversals that occurred during the study for which the ranking was assigned.

The low proportion of rank reversals lends support to the hypothesis that a stable hierarchy exists among langur females—at least in the short term. Percentages of reversals among the Abu langurs are, however, generally higher than the low percentages of reversals recorded for macaques and baboons, species believed to have highly stable female hierarchies. For example, among free-ranging rhesus macaques in a large troop on Cayo Santiago Island, Missakian (1972) recorded 61 reversals in 5,159 female-female interactions over a 24-month period—about one percent. Hausfater (1974)

TABLE 6.2. Displacements among Toad Rock troop females recorded between December 1973 and January 1974.

| Rank | Females | Displacements of ranked females | | | | | | | | | | | | Displacements of unranked females | | | Total displacements |
		Mole	Mopsa	I.E.	T.T.	Scrapetail	P.M.	No. 4	C.E.	Hauncha	Handy	Pandy	Pawlet	Niza	N-M	M-P	
1	Mole	—	10	4	2	5	1	0	3	2	0	2	2	4	0	0	35
2	Mopsa	0	—	6	6	5	2	1	0	5	0	5	3	1	0	0	34
3	I.E.	0	0	—	4	4	3	0	1	2	2	0	8	0	0	0	24
4	T.T.	0	0	1	—	2	5	0	3	0	0	0	2	2	1	0	15
5	Scrapetail	0	1	0	0	—	0	0	0	1	0	0	3	1	0	0	7
6	P.M.	0	0	0	1	0	—	0	2	0	1	0	2	0	0	0	6
7	No. 4	0	0	0	0	0	1	—	0	1	1	0	0	0	0	0	2
8	C.E.	0	0	0	0	0	0	0	—	0	0	0	0	1	0	0	2
9	Hauncha	0	0	0	0	0	0	0	0	—	3	3	0	0	0	0	6
10.5	Handy	0	0	0	0	0	0	0	0	0	—	0	0	1	0	0	1
10.5	Pandy	0	0	0	0	0	0	0	0	0	0	—	0	0	0	0	0
12	Pawlet	0	0	0	0	0	0	0	0	0	0	0	—	0	0	0	0
	Total number of times displaced:	0	11	11	13	16	12	1	9	11	7	10	20	10	1	0	132

TABLE 6.3. Displacements among Toad Rock troop females recorded between April and June 1975.

Rank	Females	Displacements of ranked females													Displacements of unranked females			Total displacements
		Hauncha	Pandy	Handy	Scrapetail	Pawlet	Mole	P-M	No. 4	T.T.	I.E.	Niza	N-M	C.E.	Moli	Vert	Infant-2	
1	Hauncha	—	5	6	0	5	0	5	1	4	4	10	8	1	1	0	1	51
2	Pandy	0	—	2	2	0	0	0	2	4	1	4	1	3	1	1	7	28
3	Handy	0	0	—	2	0	0	0	5	2	0	1	0	4	0	0	0	14
4	Scrapetail	0	0	0	—	2	2	1	1	0	0	0	0	1	0	0	1	8
5	Pawlet	0	0	0	0	—	3	2	2	5	4	2	2	6	0	0	0	26
6	Mole	0	0	0	1	1	—	1	0	2	1	0	0	0	0	0	0	6
7	P-M	0	1	0	0	0	0	—	0	0	1	2	5	1	3	0	1	14
8	No. 4	0	0	2	0	1	0	0	—	0	0	5	2	0	3	0	0	13
9	T.T.	0	0	0	0	0	0	0	0	—	1	0	0	1	0	0	0	2
10	I.E.	0	0	0	0	0	0	0	0	0	—	1	2	2	2	0	0	7
11	Niza	0	0	0	0	0	0	0	0	0	0	—	2	0	1	0	0	3
12	N-M	0	0	0	0	0	0	0	0	0	0	0	—	0	0	0	0	0
13	C.E.	0	0	0	0	0	0	0	0	0	0	0	0	—	0	0	0	0
	Total number of times displaced:	0	6	10	5	9	5	9	11	17	12	25	22	19	11	1	10	172

reported 6 reversals in 1,638 displacements recorded for females of all ages in a troop of Amboseli baboons (0.4 percent) over a 14-month period. According to both workers, female hierarchies in these groups are more stable *over time* than those of males. In the case of langurs, however, neither sex appears to have a hierarchy that is stable over many years. The average reign of an alpha male at Abu was a little over two years, that of an alpha female less than two—even including several long-lasting exceptions, Elfin and Bilgay.

Despite their short duration of rank orders, the outcome of displacement interactions at Abu was predictable on a day-to-day basis. It is unfortunate, therefore, that langurs have become a paradigm for a species in which females have poorly defined and nonpredictable dominance relations. There are several possible explanations for the lack of agreement among langur fieldworkers. The first is that either the population at Abu somehow differs from that at other langur study sites or that some environmental factor such as frequency of artificial provisioning is responsible for manifestation of a hierarchy that remains latent elsewhere. Second, it is possible that the populations are the same but that at Kaukori, Dharwar, and Polonnaruwa displacement data were pooled together so that changes over long periods obscured the actual short-term stability of the hierarchy. Third, there may have been differences among observers. Only after the 1972 field season was I especially attentive to female dominance interactions, and only after 1972 did I became convinced of their importance.

As shown in tables 6.2 and 6.3, the capacity of one female to displace another was fairly constant *within* periods of several months over which data were collected for Toad Rock troop, but not necessarily constant *between* study periods, which were separated by gaps of up to fourteen months. In two of three troops studied (the Bazaar and Toad Rock troops) major changes occurred. In no case was the rank order correlation between years statistically significant. However, the case of the Hillside troop—the smallest and, in terms of the female hierarchy, the most stable of the three troops—the correlation observed between rank orders for 1972 and 1973 could have occurred by chance only one out of ten times (Blaffer Hrdy, 1975).

Changes over time for the rank ordering of females in Toad Rock troop are shown in table 6.4 and summarized in fig. 6.9. Hierarchies for two or more years are juxtaposed, and the age and reproductive status of the female during that study period indicated. In

TABLE 6.4. Changes in approximate rank order, age, and reproductive status of Toad Rock troop females between the 1972 and 1975 study periods.

June–September 1972				February–March 1973				December–January 1973–74				April–June 1975				
Rank	Female[a]	Age	Reproductive status[b]	Rank	Female	Age	Reproductive status	Rank	Female	Age	Reproductive status	Rank	Female	Age	Reproductive status	Weight (kg)
1	Pawlet	Middle-age	Cycling	1	Mopsa	Middle-age	Estrus[c]	1	Mole	Young adult	Pregnant/infant-1	1	Hauncha	Subadult	Unknown[d]	9.5
				2	Mole	Young adult	Cycling	2	Mopsa	Middle-age	Pregnant/infant-1	2	Pandy	Subadult	Unknown[d]	9.1
								3	I.E.	Middle-age	Pregnant	3	Handy	Subadult	Unknown[d]	—
								4	T.T.	Middle-age	Pregnant/estrus	4	Scrapetail	Young adult	Unknown[d]	—
								5	Scrapetail	Young adult	Infant-2/cycling	5	Pawlet	Middle-age	Cycling	11.3
								6	P-M	Middle-age	Infant-2/estrus	6	Mole	Young to middle-age	Cycling?	11.8
								7	No. 4	Juvenile	Immature	7	P-M	Middle-age to old	Weaning	11.8
								8	C.E.	Old	Unknown	8	No. 4	Subadult	Unknown[d]	7.7
								9	Hauncha	Juvenile	Immature	9	T.T.	Middle-age	Pregnant	—
								10.5	Handy	Juvenile	Immature	10	I.E.	Middle-age	Weaning/pregnant	11.6
								10.5	Pandy	Juvenile	Immature	11	Niza	Juvenile	Immature	6.8
								12.	Pawlet	Middle-age	Infant-2/cycling?	12	N-M	Juvenile	Immature	5.9
												13	C.E.	Old	Weaning	—

a. Information insufficient to rank.

b. Cycling indicates that menstrual blood in addition to periodic estrous behavior was observed. Cycling? indicates that, based on observed estrous behavior, it was suspected that the female was cycling, but that menstrual blood was never detected nor was any subsequent pregnancy recorded. If only estrus is listed, the female was observed to solicit males but such solicitations were mild or else did not follow a cyclical pattern. Infant-1 designates a female accompanied by a newborn still wearing the dark natal coat. Infant-2 designates a female accompanied by an unweaned infant older than five months. Weaning designates a mother weaning an infant but not otherwise classifiable. Pregnant/infant-1 indicates that the female gave birth during the study period.

c. It was suspected, but never verified that Mopsa was pregnant at this time. If Mopsa did give birth subsequent to March, 1973, this infant must have died prior to about July, 1973, in time for Mopsa to conceive her next infant, born January 9, 1974.

d. Suspected of being in the early stages of pregnancy.

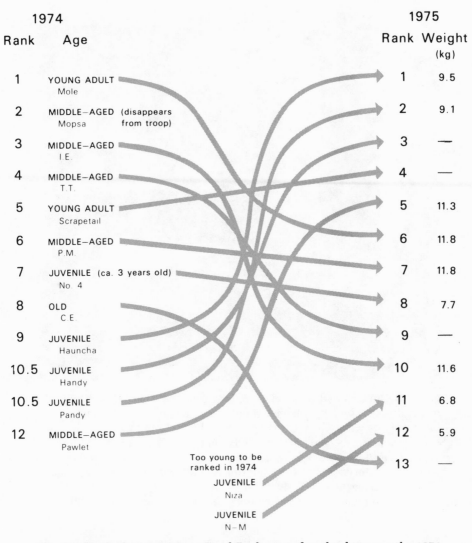

Fig. 6.9. Rank changes among Toad Rock troop females between the 1974 and 1975 study periods (summarized from table 6.4).

1975, the weight of some females was also known. Animals were weighed as shown in fig. 6.10 and further described in appendix 2.

In two troops, the Hillside (table 6.5) and Bazaar troops (table 6.6), the alpha female maintained her preeminent position through all study periods for which information was available. By contrast, in the third, Toad Rock troop, the alpha position was characterized

Fig. 6.10. During the drought of 1975, langurs were weighed to the nearest one-half pound by luring them onto a string bed suspended from a scale. Most weights were taken between 7:00 and 8:30 in the morning, but in a few cases animals could not be lured onto the scale until later in the day. When several weights taken from the same animal on the same day were available, differences between morning and afternoon weights were never greater than one pound. Weights taken on different days could differ by as much as 2.5 pounds or approximately 1 kilogram (see appendix 2).

by dramatic changes. During the 1972 study period, Pawlet, the middle-aged female with a deformed right hand, was able to displace every other animal in the troop except Splitear, the alpha male. Pawlet was an unusually aggressive female, both in terms of the frequency with which she would threaten or attack other animals and her willingness to charge human observers. Though the identification of other Toad Rock females was not reliable in 1972, this female with a deformed paw was unmistakable. On more than fifty occasions, Pawlet displaced other animals. She herself was

TABLE 6.5. Changes in approximate rank order, age, and reproductive status of Hillside troop females between the 1972 and 1975 study periods.[a]

	June-September 1972				February-March 1973				April-June 1975		
Rank	Female	Age	Reproductive status		Rank	Female	Age	Reproductive status	Rank	Female	Age	Reproductive status
1	Bilgay	Young adult	Infant-1/ cycling		1	Bilgay	Young adult	Pregnant	1	Bilgay	Young adult– middle-age	Pregnant/ infant-1
2	Itch	Young adult	Infant-1/ cycling		2	Oedipa	Middle-age	Estrus	2	Itch	Young adult– middle-age	Weaning
3	Harrieta	Middle-age	Infant-2/ estrus		3	Itch	Young adult	Estrus	3	Oedipa	Middle-age	Infant-2
4	Oedipa	Middle-age	Infant-2		4	Harrieta	Middle-age	Infant-2	4	Harrieta	Middle-age	Pregnant/ infant-1
5	Pawless	Middle- age–old	Cycling?		5	Pawless	Middle- age–old	Infant-1	5	Pawless	Middle- age–old	Infant-1
6	Sol	Old	Unknown		6	Sol[b]	Old	Cycling				

a. Information was not sufficient to rank all females for December-January 1973-74. During this study season, Bilgay retained the alpha position while Pawless and Sol remained the two lowest ranking females. It was not possible to rank Oedipa, Itch, and Harrieta in relation to one another.

b. Sol was rarely present with the troop; her low rank is only assumed on the basis of her rank the previous year and the fact that she would compete with no other female.

TABLE 6.6. Changes in approximate rank order, age, and reproductive status of Bazaar troop females between the 1973 and 1975 study periods.

1973				1973-74				1975				
Rank	Female	Age	Reproductive status	Rank	Female	Age	Reproductive status	Rank	Female	Age	Reproductive status	Weight (kg)
1	Elfin	Young adult	Cycling	1	Elfin	Young adult	Infant-1	1	Elfin	Young adult	Pregnant	12.3
2	Nony	Young adult	Pregnant	2	Kasturbia	Young adult	Infant-2	2	Nony	Young adult	Infant-1	—
3	Overcast	Young adult-middle-age	Cycling	3	Breva	Young adult	Pregnant	3	Breva	Young adult	Weaning/estrus	10.5
4	Kasturbia	Young adult	Infant-1/cycling?	4	Pout	Middle-age	Weaning	4	Pout	Middle-age	Infant-1	—
5	Earthmoll	Middle-age	Infant-2/cycling	5	Junebug	Subadult/young adult	Cycling	5	Junebug	Young adult	Pregnant	8.2
6	Pout	Middle-age	Infant-1	7.5	Nony	Young adult	Infant-2	6	Guaca	Young adult	Weaning	8.4
7	Quebrado	Old	Unknown	7.5	Overcast	Middle-age	Unknown	7	Kasturbia	Middle-age	Pregnant	10.9
8.5	Breva	Subadult	Cycling	7.5	Guaca	Young adult	Pregnant/infant-1	8	Overcast	Middle-age	Cycling	10.5
8.5	Short	Old	Unknown	7.5	Short	Old	Estrus	9	Nullipara #2	Subadult	Estrus	6.8
10	Wolf	Middle-age-old	Infant-1	10	Earthmoll	Middle-age	Infant-2/estrus	10	Nullipara #1	Subadult	Estrus	6.4
11	Junebug	Subadult	Estrus	11	Quebrado	Old	Estrus	11	Earthmoll	Middle-age	Pregnant	—
12	Guaca	Subadult	Cycling	12	Wolf	Middle-age-old	Loses infant/resumes cycling	12	Short	Old	Unknown	11.1
								13	Quebrado	Old	Unknown	10.9
								14	Wolf	Middle-age-old	Pregnant	—

never displaced except by Splitear, who on occasion would fiercely attack but never wound her.

In sharp contrast to this 1972 season, during the 1973-74 season, Pawlet was displaced more than twenty times, but was never seen to threaten or displace any other animal. Pawlet's plunge in rank was already apparent in the early months of 1973, when another middle-aged female called Mopsa had become the alpha female. Mopsa had ranked somewhere just below Pawlet in 1972.

By early 1973, Mole, the young female with a sleek silver-white coat, had moved into the beta position just under Mopsa. When re-encountered in the 1973-74 study period, Mole unquestionably filled the alpha position. She displaced Mopsa on at least ten occasions and was never in turn displaced by her. In most instances, Mopsa avoided behavior (such as eating in the proximity of Mole) that might elicit threats from the younger female.

During Pawlet's 1972 period of supremacy, there may have existed a special relationship between this female and the alpha male, characterized by close association at some times and by fierce "cat-fighting" at others. During Pawlet's estrous periods, she would lunge at, slap, and chase Splitear whenever he approached. Splitear retaliated by chasing her (figs. 6.11 a, b). After Toad Rock troop was usurped by Toad in 1973, no special relationships between any female and the new male were detected.

It is not clear whether or not Pawlet's aggressiveness was related to sexual cycling. When considering her antagonism towards Split-ear during estrus, one should keep in mind that Splitear was probably more likely to encroach on Pawlet's personal space at such times. Nevertheless, langur females, like some other primate species, show a pronounced increase in threat behaviors and involvment in fights during estrous behavior. Sol, Overcast, Junebug, and probably others, in addition to Pawlet, exhibited particularly aggressive behavior during one or more estrous periods (see Hall and DeVore, 1965, for baboons; Yerkes, 1939, for chimpanzees; Bernstein, 1963, for rhesus; and Oakes, cited in Bardwick, 1974, for increased assertiveness among human females at midcycle). In a number of cases, I could correctly predict that a female was in estrus even before she solicited a male on the basis of her irritability. Estrous females were more likely to threaten, lunge at, or slap other langurs and human observers. The only two injuries I ever sustained from langurs were scratches inflicted by estrous females. On a third occasion, I was so startled by a sudden charge from an estrous Pawless (whom I had not been watching) that I stepped back-

Fig. 6.11. During the days when Pawlet was in estrus (a), relations between this alpha female and the troop leader (Splitear) were characterized by fierce cat-fights (b).

wards into a crevice. It was my impression that the presence of one or more estrous females dramatically increased the frequency of threats, displacements, and redirected aggression within the troop, but an increased frequency of displacements does not necessarily mean a change in the social structure.

None of the rank changes that were recorded over time in these three troops could have been predicted on the basis of reproductive state. Bilgay, for example, passed through various reproductive stages, including weaning an infant, cycling (in 1972), pregnancy (in 1973), and bearing new infants (in 1973 and 1975) without any apparent effect on her ability and willingness to displace every other female in the Hillside troop. The same was true for Bazaar troop's alpha, Elfin, for the three seasons during which information was available on her status.

The direction of rank change, however, could be fairly well—but by no means completely—predicted on the basis of age. Whereas younger females tended to rise in the hierarchy, aging females tended to fall in rank. In Bazaar troop between February and March 1973 and December and January 1973-74, 3-to-4-year-old Guaca rose from twelfth to sixth place, Junebug from eleventh to fifth place, and Breva from tied-for-eighth to third place (fig. 6.12). These young females began to rise toward the top portion of the hierarchy months before giving birth to their first infant (Guaca in January 1974; Breva in February 1975; Junebug in May 1975). Conversely, during the same period, Bazaar troop's Earthmoll, Overcast, and Wolf, all middle-aged or older females, slipped downwards in rank. The fact that Hillside troop was relatively stable over the years of this study may reflect the fact that no young females in Hillside troop survived to grow up and move up in the hierarchy.

During the 1975 study period, I had the opportunity to test explicit predictions based on the hypothesis that young females tend to rise in the hierarchy at the expense of aging ones. Just before returning to Mount Abu for the fifth time, I ventured the following forecast for Toad Rock females: No. 4, Hauncha, Handy, and Pandy would be expected to rise in rank, while Mopsa, I.E., and T.T. should become less able or likely to displace other females (1975:223). As shown in table 6.4, Mopsa had disappeared from the troop, T.T. had fallen to ninth place, I.E. to tenth. At the top of the hierarchy was Hauncha, a 9.5 kilogram (21 pound), not quite four-year-old female who had yet to reach full body size. Pandy and Handy ranked just below her, with No. 4 in eighth place (fig. 6.4). That the prediction should come true was not surprising, but the speed with which these extreme shifts transpired exceeded my expectations.

Despite the predictive power of age in these cases, it is unlikely that age will turn out to be the only factor determining rank. Pawlet's fall in rank and subsequent rise, for example, cannot be adequately explained by age alone, and to date remains a mystery. In

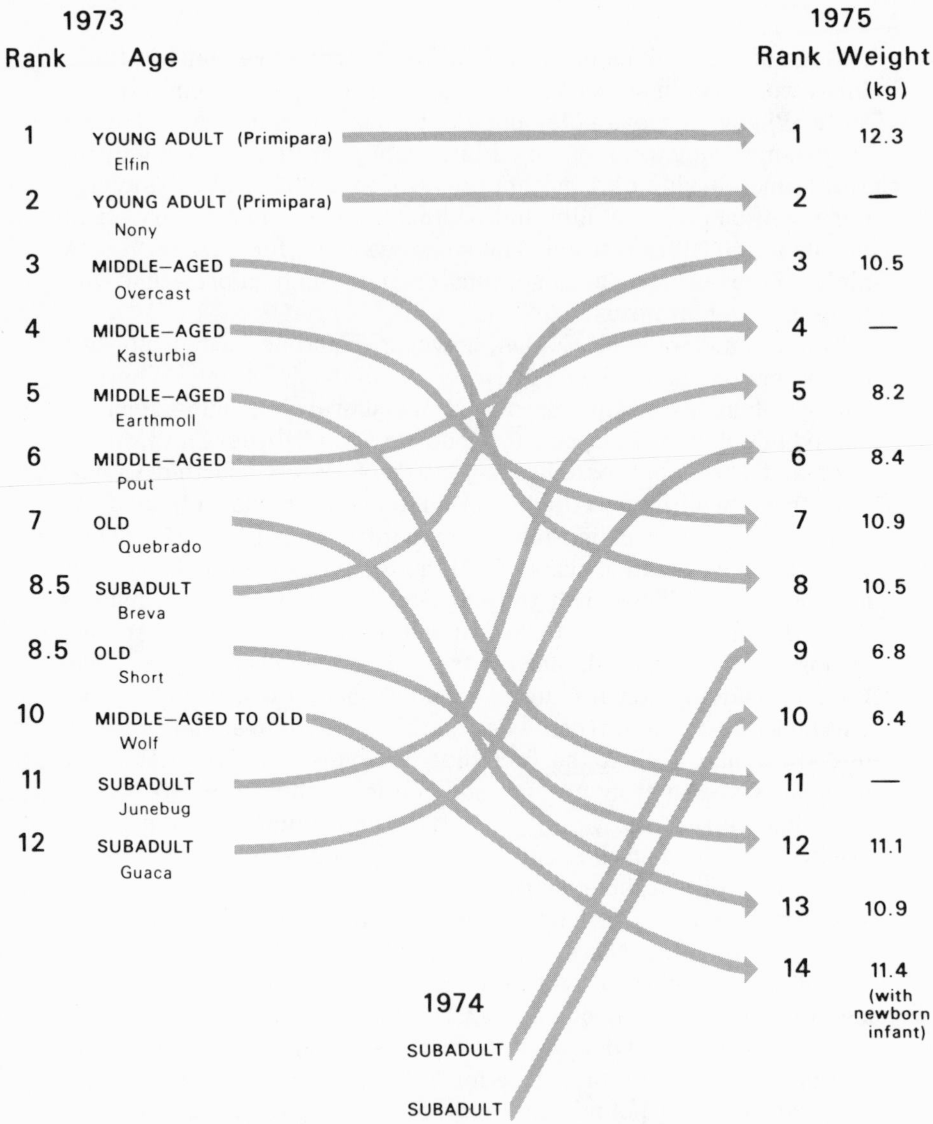

Fig. 6.12. Rank changes among Bazaar troop females between the 1973 and 1975 study periods (summarized from table 6.6.).

some cases factors such as physical handicaps may be important. Wolf and Overcast, who were slipping in rank, were not only aging but each blind in one eye. Although the three-legged female Guaca was the oldest female in her age-group, and hence under normal conditions might have begun her rise in rank sooner than Junebug (who was a year younger) and Breva (six months younger), she was consistently subordinate to both these younger females throughout subadulthood.

I am now convinced that other things being equal, young adult and subadult females tend to rise in rank at the expense of their elders, sojourning during their prime in the top portion of the hierarchy. However, it should be noted that this generalization was derived from limited observations of only three troops whose genealogies were largely unknown under the specific conditions which prevailed at Abu between June 1971 and 1975. It has too long been a practice of primatologists to extrapolate from a single study site to a whole species. Until more extensive—and more rigorous—information about female hierarchies is available for langurs in other areas, generalizations encompassing the entire species would be premature. Nevertheless, a close reading of two doctoral theses on the social organization of langurs at Kaukori and Polonnaruwa provides support for some statements made here. At first glance the statement that young females rise in rank over older ones is not in agreement with Ripley's observations at Polonnaruwa. In fact, Ripley specifically states that "dominance relations between females are difficult to determine . . . [but in general] older multiparous females are more dominant than younger primiparous or nulliparous adult" (1965:289). But this generalization is contradicted by Ripley's diagram of Polonnaruwa troop structure (table 25). Two of the highest ranking females (Solo and Blank) are listed elsewhere as being young adults, both presumed to be primiparous (pp. 71-72). Although there was no consistent female hierarchy at Kaukori, Jay (1963c) noticed that old females almost never displaced other females in the troop; furthermore, the oldest female in Jay's Kaukori troop (female R) spent "approximately 80 per cent of her day . . . on the extreme edge or even outside of the troop. Although she slept and travelled with the troop she rarely took part in grooming groups or in any other activity which brought her in close contact with other adults" (p. 141). Similarly, female F, the oldest female in the Kaukori troop associated with an infant and probably the second oldest female in Jay's study troop, was "nervous when sitting close to very dominant animals and whenever possible she avoided being with large groups of adults for more than 10 or 15 minutes. F stayed near the edge of the troop for well over one-half of the day rather than enter grooming groups with other adult females. Although less than 10 per cent of her threats resulted in displacements of other females she was usually irritable and displayed mild threat gestures such as slapping the ground more frequently than any other adult female" (p. 137).

The avoidance of troop members by females R and F at Kaukori exactly parallels the peripheralization of old females Sol, Short, Quebrado, C. E., and to some extent Wolf, at Mount Abu. Each was among the oldest females in her respective troop, and each spent long periods on the periphery or completely apart from her troop. By 1974, the year before she died, old Sol of Hillside troop spent whole days alone. In her 1963 report, Jay noted the peculiar irritability of old females in her Kaukori study troop, Similarly at Abu, old females would threaten, charge, or slap a human harasser on little provocation.

Whereas young females intrepidly initiate contacts with other langurs, actively approaching, grooming, embracing, or competing with them, old females to a remarkable degree avoid the close company of their fellows. A few hundred minutes taken from the lives of two extreme cases, a not-quite-four-year-old-top-ranking nulliparous female (Hauncha) and an old, low-ranking multipara (C. E.), viewed during the 1975 drought, illustrates behavioral differences typical of young and old females at Mount Abu.

During those morning and afternoon hours in 1975 when members of the Toad Rock troop were most active, clearly visible, and undisturbed by outside events, I systematically focused on specific females in this troop and noted down every social contact the focal female had during a certain period of time. For example, in 301 minutes of observation spread over several days, Hauncha approached other troop members on 28 occasions, displacing them on 19 of these. In 12 such displacements, Hauncha overtly threatened the animals she displaced by lunging at them. Hauncha was never displaced by any other animal except the alpha male and was approached by others on only five occasions. Additionally, on nine occasions Hauncha approached animals (usually juvenile males) in the neighboring School troop to threaten, chase, play-wrestle, or groom them.

Curiously, even though Hauncha was able to prevail over every other female in the troop, she was by no means the largest or heaviest. Dominant animals typically mount their subordinates, but in Hauncha's case, mounting was complicated by the fact that she was smaller than some of her full-grown subordinates, such as P-M. P-M was taller, broader, and heavier, weighing 11.8 kilograms (26 pounds) compared to 9.5 kilograms (21 pounds) for Hauncha. In order to mount P-M, Hauncha would climb up onto P-M's back, her chin against P-M's shoulder, clinging there like a terrified equestrian.

In contrast, during 478 minutes that C.E. was watched, I saw her approach another langur only once—a cream-colored infant that C. E. caught hold of by the rump. She never initiated grooming, hugs, or mounts (table 6.7). Both spatially and temporally, old C.E. avoided other animals. When she fed, C.E. would often forage on the outskirts of the troop or else absent herself from it entirely. If the main food source happened to be occupied by other members of the troop, C.E. would do her feeding while other members dozed or were absorbed in grooming activities. When C.E. did attempt to feed in company with others, she fared poorly. On three occasions C. E. was threatened from an advantageous position at a palm tree oozing sap, twice from a scavenging position at a garbage heap. On five occasions, C.E. was actually attacked and contacted by another female bent on usurping C.E.'s food source. At the conclu-

TABLE 6.7. Contrast in their involvement in social interactions between a high-ranking young female (Hauncha) and a low-ranking, old one (C.E.).[a]

Interaction	Hauncha		C.E.	
	Number of occasions	Number of occasions per 100 minutes	Number of occasions	Number of occasions per 100 minutes
Approaches another langur	28	9	1	0.2
Displaces another	19	6	0	0
Grooms another	10[b]	3	0	0
Embraces another	3	1	0	0
Mounts another	6	2	0	0
Play-wrestles	11	4	0	0
Approached by another langur	5	2	21	4
Displaced by another female	0	0	19	4
Groomed by another	7	2	5	1
Embraced by another	1	3	1	0.2
Mounted by another	0	0	1	0.2

a. Total number of minutes of focal-female sampling: Hauncha = 301, C.E. = 478.
b. Sixty percent of this grooming was directed towards Toad.

sion of one such attack by Mole the fur on C.E.'s shoulder was specked with blood. In these cases, C.E. would immediately cease feeding and move away chattering her lips (an unusual behavior for langurs). On two occasions, other females (Pawlet once, and Pawlet's daughter, Niza, once) attempted to remove food hanging from C.E.'s mouth, but did so without displacing her.

The lives of old female langurs, then, are not altogether enviable: Sol and Pawless risk their well-being to protect another female's infant while the mother herself remains on the sidelines; Short and Quebrado repulse females from other troops but do not share in the spoils of their success; old C.E. is displaced by almost every other animal in the troop except infants. Their plights are typical of old females. It is these hapless oldsters who hold the key to understanding how the langur dominance system works and how the troop's resources are apportioned. It took me four fieldtrips to arrive at this conclusion and a fifth to return to Abu to test the predictions which flowed from it.

Correlates of Female Rank

Fitness. Maximal strategies for a female include insuring that any infant she produces survives, and that the interval between births is as small as is compatible with infant survival. Though there is a ceiling to the number of offspring that a female can produce in a lifetime, this maximum number is rarely achieved under natural (less than ideal) conditions. One route toward this maximum reproductive potential is for a female to increase her access to resources that can be converted by her into offspring, that is, through competition with other individuals. Presumably, a female's success in such competition can be measured by her rank. The question remains, however, to what extent this rank is actually correlated with fitness.

For many species such as pig-tailed and rhesus macaques, hierarchies are stable over time (Bernstein, 1969a; Missakian, 1972). Among Japanese and rhesus macaques for whom longitudinal information over several generations is available, regular patterns of ranking emerge (Sade, 1972). Daughters routinely fit into the hierarchy just under their mothers, with younger sisters ranking just above their elder sister. Typically, the rank of these matrilineages is fixed relative to one another. However, when hierarchies are disrupted either through external or internal events, females as well as males have been observed to engage in fierce and, in extreme cases, murderous fights. During a cage study of the pig-tail macaque, a young male's bid for top rank resulted in the death of a high-ranking female who opposed his rise (Bernstein, 1969b; see also Rosenblum, 1971). Similarly Marsden (1968) reports that when a dominant male was experimentally replaced among a caged group of rhesus, fighting broke out between two high-ranking females and their offspring to determine which lineage would be dominant. Fights resulting in murder or near-murder have also been reported for free-ranging macaques in India and Japan (Lindburg, 1971; J. Kurland, personal communication).

Whereas competition between males can be explained by the relative investment of both sexes and the high yield in fitness for the victorious male (Trivers, 1972), the explanation for why females would fight so fiercely that death sometimes ensues is less clear. In short-range terms, high rank may mean access to such priorities as preferred or limited food resources, a position near to dominant males and other species-specific perquisites of rank such as being groomed (Simpson, 1973, for chimpanzees). The best documentation to date of precisely how many more grains of wheat high-and low-ranking monkeys get in a free-ranging provisioned situation comes from a food-intake study of Japanese macaques on Koshima Islet (Iwamoto, 1974). Iwamoto found that the top-ranking five females in a troop of twenty-three females ate on average just under 5,000 grains of wheat. The remaining females averaged about 2,000 grains apiece. In this study group, the highest-ranking lineage uniformly obtained more grain.

In long-range terms, high rank may mean greater fertility or a greater number of offspring that survive. Until recently, evidence to support a correlation between a female's rank and her reproductive success was largely anecdotal. Barbara Smuts (1972) reported that among the rhesus females she observed on La Parguera Island, the highest ranking appeared to leave more surviving offspring. Thelma Rowell noticed that in the breeding colonies she set up for six distinct species of Old World monkeys, "in all cases the highest-ranking were the first to breed, and with few exceptions females became pregnant in order of descending rank. Higher-ranking females also tend to end their lactation interval sooner" (1972a).

Rowell also noted that "when females were beaten up by other females their menstrual cycles were lengthened" (1970). A similar phenomenon occurs among savanna baboons: harassed females temporarily lost their estrous swelling (Gillman and Gilbert, 1946). It is possible that females may be suppressing estrus in their fellows by stessing them as well as by displacing them for resources.

It was not until 1974 that Lee Drickamer published firm evidence for the correlation between female rank and reproductive success based on records kept for the rhesus population on La Parguera Island over a ten-year period. These population statistics indicate first that a larger percentage of high-ranking females breed each year; that offspring of such females have a higher rate of survival than do infants born to lower-ranking females and that the daughters of high- and middle-ranking females themselves produce more offspring at a significantly earlier age (3.8 – 3.9 years) than do daughters of low-ranking females (4.4 years). Among rhesus macaques, then, a mother's rank has a potentially great effect on her genetic contribution to subsequent generations. Under these circumstances a female's willingness to compete for high rank could well be worth the potential cost to her from fighting.

No comparable data on female reproductive success exist for langurs. Furthermore, if young females are moving up the hierarchy at the expense of aging ones, all females might pass through more or less the same rank positions in the course of a lifetime. It is difficult to see how rank-related differences in overall reproductive success could arise in this hypothesized system of a "rotating chairmanship." Nevertheless, it is possible that a given female might increase her fitness by consistently attaining a higher rank over her peers at each age stage or by consistently prolonging her stay in a position of privilege. Such prolongations could occur either as a result of historical accident (the case of Bilgay who remained alpha female of Hillside troop year after year during a period when no younger animals survived), as a consequence of personal traits (Bazaar troop's Elfin may illustrate such an example), or, possibly, genealogy. Conceivably, then, the nepotism, or nonequal lifetime benefits determined by genealogy, that is so apparent in the macaque system could have evolved from a system similar to the one hypothesized for langurs. Nevertheless, there are some obvious but crucial differences between the two systems. In a nepotistic system, fitness differences between females occupying different ranks affect generation after generation and over time could be potentially very great; competition should be fiercer than in a system where high rank is impermanent. Furthermore, the competition might be intensified among species such as macaques because of the lower degrees of relatedness among members of troops in which there is typically more than one adult male breeding at the same time. Interestingly, this greater commitment to intratroop dominance relations among nepotistic macaque species could explain the domination of langurs by single macaques in those instances where isolated macaques have joined langur troops. In the langur system in which females rotate through a succession of ranks—much as males do in a number of primate species including langurs and baboons (Hausfater, 1974)—females, rather than inheriting a fixed position, would in general be more individualistic in their dominance dealings.

The near absence of alliance-type relations among langur females fits well with this suggestion. In contrast to species such as savanna baboons, hamadryas baboons, and macaques where defensive alliances are a way of life (Hall and DeVore, 1965; Kummer, 1968; Bernstein, 1969a), at Abu I witnessed only one instance in which the presence of a second female seemed to significantly alter the position of a third party (Blaffer Hrdy, 1975). On this occasion, it appeared to me that a subadult female (Breva) was able to displace a female who was normally dominant to her (Elfin) because of the proximity of a third adult female (Kasturbia). In all other cases that included the defense of a third animal such as an infant by two or more females, it was not possible to distinguish the "alliance" from joint attacks brought about by simultaneous motivation in more than one animal.

A consequence of a nepotistic system where rank is inherited from the mother is that males should be more selective in their mate choice, whenever they have a choice and whenever there is a limit (for example, to time or energy) or a cost attached to females inseminated (when fighting would be necessary, for example). Only among species where there is a significant correlation between female rank and reproductive success would males be expected to exhibit any strong bias for higher-ranking females. Differing degrees of reproductive fitness between females of different ages might also influence male preferences. For example, increased survival of infants born to older females could provide an explanation for the tendency of more dominant males to associate preferentially with older, usually high-ranking multiparous females while ignoring the solicitations of young females among such varied species as wild and free-ranging rhesus macaques (Conaway and Koford, 1965; Bernstein and Sharpe, 1966; Hanby, Robertson, and Phoenix, 1971; Lindburg, 1971), anubis baboons (Ransom and Ransom, 1971), and perhaps chimpanzees (the remarkably popular old female Flo described in van Lawick-Goodall, 1971). This interpretation fits with data on infant survival and maternal age recorded for rhesus macaques at La Parguera Island: whereas 40-45 percent of first- and second-born infants died prior to weaning, only 21 percent of infants born to females seven years or older died (Drickamer, 1974). According to Conaway and Koford, "The tendency of older females to be most frequently associated with the dominant male may restrict the number of females bred by him so that he is less effective than some lower-ranking males in siring offspring . . . Thus much of the breeding effort of the dominant male may be wasted on comparatively few females." In fact, however, in terms of the number of his offspring that actually survive, the preference of a dominant macaque or baboon male for an older female may be anything but a waste of time.

In contrast to the poor reproductive efficiency of young rhesus mothers, there is no evidence that langurs lose such a disproportionate number of first- and second-born offspring. The greatest single cause of mortality at Abu was infanticide, and primiparas do not fare substantially worse in this respect than older mothers. A breakdown of known infant mortality at Abu into age class of the mother indicates that of three infants belonging to old mothers, one died prior to weaning. Twenty-nine percent of twenty-one infants born to twelve middle-aged-to-old mothers died, while 38 percent of sixteen infants born to eight young adults died. Once a mother loses her offspring, however, there does seem to be a difference in how quickly females of different ages can recoup their loss and produce a new offspring. To take one example, Bilgay, a young-to-middle-aged female in Hillside troop, and Pawless, a middle-aged-to-old female of the same troop, both lost half of the infants they were known to produce between 1971 and 1975. The younger female, however, was able to produce at least four infants in the course of the study while Pawless was only seen with two. On average, females classed as old, middle-aged, and young produced 0.18, 0.38, and 0.67 infants per year, respectively. (Note though that the rate of production for young females is biased upward by a system which starts counting offspring as soon as a female gives birth.) By any standards, rate of reproduction by old females was low. Of the four oldest animals at Abu, two (C.E. and Short) produced one infant apiece in the four years in which they were identified, Quebrado produced one in five, while Sol was never seen with an infant in any of the four study periods up to her death.

These findings suggest that langurs and rhesus macaques have quite different schedules of reproductive fitness, with macaques peaking in fitness later than langurs (at about seven years).[4] "Reproductive fitness" here differs from Fisher's classic definition of reproductive value, which does not take infant survivorship into account. Fisher's reproductive value is a simple measure of how many offspring a female of a given age is likely to produce in the years remaining to her regardless of whether or not those offspring survive. The curve of this function is approximately the same for both macaques and langurs: peaking at sexual maturity and declining with age.

Important differences emerge from this comparison of macaques with langurs. Nepotism and the late timing of macaque reproductive fitness create powerful biases for the reproductive success of different females and may also affect male choice of mates. The stage is set for fierce competition between macaque females, not directly for males, but for status. In addition, the lower degrees of relatedness within macaque troops in which several males are simultaneously breeding devalues cooperation with troop members at large, while placing a bonus on cooperation with lineage mates against other matrilines. Comparable pressures favoring interindividual alliances do not exist in the small, one-male langur troops. The result is two apparently very different social systems, with quite different adaptations for female ranking. Whereas macaque females inherit their rank from their mothers, fitting into the hierarchy just below their mother and cooperating with lineage mates throughout life to retain this position, langurs inherit no such fixed position. It is not known yet what role genealogy plays in rank determination, but it is already clear that a langur's rank is not fixed at birth and that her position will fluctuate throughout her lifetime. Furthermore, it is obvious that daughters are not necessarily entering the hierarchy just below their mothers, since in some instances they rank above them.

Leadership. The only relationship between female rank and leadership was a possibly inverse one, since old, low-ranking females are more likely than younger ones to determine the direction of troop movement. In almost every instance where a female led the troop, she belonged to the lower portion of the displacement hierarchy (Blaffer Hrdy, 1975). One explanation for the apparent association between low rank and leadership is that older females are low-ranking and also happen to be the animals most familiar with the local habitat. It is also true that older females are generally

4. Calculated by Peter Ellison from data summarized in Drickamer (1974).

more likely to strike out on their own and to spend long periods foraging apart from the troop. Similarly, except in cases when the alpha male intervened to protect offspring in his harem, the role of troop defender usually fell to older females.

Sexual harassment. Likewise, factors other than rank (kinship, reproductive state, dependency states, and so on) appear to determine which animals engage in this behavior.

Mounting behavior most clearly paralleled the direction of displacements among female langurs. On more than 45 occasions females in the Bazaar, Toad Rock, and Hillside troops were observed to mount other females in a close approximation of the male copulatory position.[5] About one-third of the time, pelvic thrusting was observed. In 37 of these cases, relative rank of the partners was known. In every instance, the mounted female was subordinate. In 17 of the 37 cases, the female mounted was in estrus. In several of these, the estrous female gave the impression that she wished to avoid the attentions of other females. Such a female would curve her back so that her side faced whichever female was trying to mount her and aim her soliciting rump directly at the alpha male before shuddering her head. On three occasions when an estrous female was mounted by another female, the "copulating" couple was harassed by other adult females and juveniles. The only exception I know of in which a subordinate female mounted an animal who might otherwise have been dominant to her occurred when an estrous female briefly mounted a young male who had temporarily invaded the School troop to steal a copulation from her (16 mm film record by D. B. Hrdy).

Embracing. In contrast to the unambiguous directionality of mounting behavior, which of two females will initiate an embrace cannot be predicted on the basis of rank. However, the social context in which an embrace occurs may to some extent predict whether a high-ranking female will embrace a low-ranking one, or vice versa. Langurs embrace each other by one or both partners clasping their arms around another langur, patting the partner's back, or pressing cheeks (fig. 6.13). Not uncommonly, one or both partners grimace. Ripley has likened this behavior to the continental "abrazo." Langur fieldworkers tend to agree that an embrace is a reassuring or placating gesture and that it reduces tensions in a species where interindividual relations are characterized by a

---

5. What appeared to be mounting of infants, that is, a bipedal female holding a very young infant between her legs, is not included in this sample.

Fig. 6.13. Female-female embrace.

great deal of tension (Sugiyama et al., 1965; Ripley, 1965; and Jay cited in Ripley).

In 36 of 48 episodes in which one female embraced another, the identity and relative rank of both partners were known. In 27 of these, it was possible to determine which female initiated the hug. No difference in the likelihood of a subordinate versus a dominant female initiating the embrace could be detected. However, the social contexts in which dominant versus subordinate females might be likely to initiate the embrace may differ. Though it should be kept in mind that the sample sizes are small and that these episodes were taken from the ad libitum record with no attempt other than common sense to preclude sampling biases, several interesting patterns emerge. First, embraces initiated by dominant females were accompanied by grooming 64 percent of the time. One possible explanation for embraces in this context is that the dominant female wishes either to groom or be groomed and she is reassuring the lower-ranking female by hugging her in order to keep the prospective grooming partner from being displaced and moving away. By contrast, on 13 occasions when a subordinate female initiated the hug, the episode was accompanied by grooming only 23 percent of the time. Thirty-one percent of the time subordinates hugged dominants in a context of social disruption, that is, just after one female had been threatened or displaced by another or when the troop was threatened by some outside agent (Blaffer Hrdy, 1975). These data, scant as they are, tantalize us with the possibility that more is at issue in embracements than mere "greeting" (see Ripley, 1965); important messages between individuals may be conveyed by this act.

Grooming is among the most common and relaxed of langur social relationships (figs. 6.14, 6.15). The tension apparent between adult individuals at almost all other times is indiscernible. As in other primates, grooming relations are complex, a subject of study in themselves (for example, Sparks, 1969; Simpson, 1973). No in-depth study of grooming was undertaken at Abu, and the results presented here are taken from the ad libitum record. From a small sample of 178 grooming interactions in which the sex of the grooming partners and the duration of the session were known, it is apparent that females groom other animals significantly longer than males do (table 6.8). Generally, adult males only groom females during consort relations. On average, males took more time to groom females than to groom other males, but this difference was not statistically significant. The longest durations were recorded for females grooming males, and these grooming stints were significantly longer than the average duration for females grooming other females.

No obvious relationship between rank and the direction of grooming was apparent. Not surprisingly, females who groomed other females were likely to be groomed themselves. As shown in table 6.9

Fig. 6.14. Middle-aged Bazaar troop female, above, grooms a young adult female. Juvenile in lower branches grooms herself.

Fig. 6.15. It is by no means clear that langurs prefer to groom close relatives; this episode lasted thirty minutes.

the middle-aged female Oedipa did the most grooming, and was in turn groomed the most by other females. It is not known whether taking into account the duration of each grooming episode would change this picture.

Access to infants. High-ranking females are relatively successful at obtaining infants, but age and experience are also important factors in determining which females will succeed or fail. Langur females are attracted to newborn infants, some females more so than others. Interested females attempt to take the infant from its mother or from another caretaker, to hold and carry it. Of 808 attempts to take infants, 196 were abortive. In all but two cases, failure could be attributed to one of the following causes: refusal of another caretaker to give up the infant (accounting for 52 percent of failures); maternal rebuff (28 percent of failures); hesitation by the prospective caretaker, or what was classified in my fieldnotes as "look and reach only" (14 percent); or resistance by the infant (6 percent).

In tables 6.10 and 6.11, females in the Toad Rock and Bazaar troops are listed in order of rank, and each female is scored according to her success rate in obtaining another female's infant, that is, the ratio of the total number of times she succeeded in taking an infant to the total number of times that she tried to take one. As these tables show, rank alone does not adequately predict a female's success rate, though in general females at the top of the hierarchy did well in obtaining infants. As an example of a female whose success did not conform with her rank, Pawlet was the

TABLE 6.8. The duration of grooming episodes classified by sex of partners.[a]

Episode	Average length (minutes)	Standard deviation	Sample size
Female grooms male	10.03	6.36	26
Female grooms female	3.49	3.50	131
Female grooms other animals of either sex (weighted mean)	4.57	4.11	157
Male grooms male	1.79	1.07	14
Male grooms female	3.44	2.31	7
Male grooms other animals of either sex (weighted mean)	2.34	1.6	21

a. Results of a Student's t test indicated that females groomed other animals significantly longer than the males did at the 0.001 level of significance. Females also groomed males significantly longer than they did other females at the 0.001 level. The difference between the amount of time males spent grooming other males and grooming females was not significant.

TABLE 6.9. Direction of grooming relationships observed among Hillside females between August 5 and September 12, 1972, taken from the ad libitum record.

Groomer[a]	Bilgay	Itch	Harrieta	Oedipa	Pawless	Sol	Total number of episodes
Bilgay	—	5	3	17	13	1	39
Itch	5	—	6	17	3	1	32
Harrieta	6	12	—	14	4	2	38
Oedipa	16	18	10	—	4	11	59
Pawless	7	6	1	4	—	2	20[b]
Sol	1	1	0	4	0	—	6
Total number of times groomed	35	42	20	56	24	17	194

a. Females are listed in order of their ability to displace other females.

b. During this same period, Pawless, who was cycling, groomed Mug and Shifty (when he visited) more frequently than did any other female.

TABLE 6.10. Success rate in obtaining infants for Toad Rock troop females, listed according to rank for the 1973-74 field season.

	1973				1973-74		
Rank	Female	Number of tries	Successful tries (%)[a]	Rank	Female	Number of tries	Successful tries (%)[a]
1	Mopsa	30	80	1	Mole	38	89
2	Mole	4	100	2	Mopsa	30	80
b	I.E.	4	75	3	I.E.	49	86
b	T.T.	2	100	4	T.T.	84	83
b	Scrapetail	2	100	5	Scrapetail	7	43
b	Pawlet	2	50	6	P-M	26	69
				7	No. 4	6[c]	67
				8	C-E	5	100
				9	Hauncha	33	79
				10	Handy	38	76
				11	Pandy	79[c]	72
				12	Pawlet	6	83
					Niza	40	10
					N-M	7	0
					Vert	27	4
	Total:	44				475	

a. These proportions are useful for comparisions between individual females and between groups, but they are not accurate as indicators of "absolute" rates of success; these proportions are based on the ad libitum as well as the focal-infant sample and are biased toward success.

b. Not ranked.

c. No. 4 closely resembled a second subadult, Pandy, in age and appearance, and these two females were often confused. In the early part of the study, attempts actually made by No. 4 were probably attributed to Pandy. Because of this confusion, the average number of attempts and rate of failure for all four subadults is preferable to individual scores.

lowest-ranking adult or subadult female in the Toad Rock troop during the 1973-74 season, yet her success rate in taking infants was the same as T.T.'s, the fourth-ranking female.

There is a somewhat more clear-cut association between maternal rank and the likelihood that attempts to take very young infants will be rebuffed. During the 1973-74 field season the mothers of the two youngest infants were the two highest-ranking females in Toad Rock troop. Mole (who rebuffed prospective caretakers in 8 of 69

TABLE 6.11. Success rate in obtaining infants for Bazaar troop females, listed according to rank for the 1973-74 field season.

		1973				1973-74[a]	
Rank	Female	Number of tries	Successful tries (%)	Rank	Female	Number of tries	Successful tries (%)
1	Elfin	5	80	1	Elfin	3	100
2	Nony	5	80	2	Kasturbia	8	100
3	Overcast	7	57	3	Breva	8	100
4	Kasturbia	0	—	4	Pout	1	100
5	Earthmoll	5	60	5	Junebug	30	97
6	Pout	3	67	6	Overcast	7	100
7	Quebrado	1	100	8	Nony	6	100
8.5	Breva	17	71	8	Guaca	10	80
8.5	Short	1	100	8	Short	1	100
10	Wolf	1	100	10.5	Earthmoll	7	100
11	Junebug	25	76	10.5	Quebrado	0	—
12	Guaca	31	68	12	Wolf	6	100
				13	Astra	0	—
				14	Nuka	0	—
				15	Nonet	1	100
	Total:	101				87	

a. The extraordinarily high success rate for females in the Bazaar troop during this season had to do with the fact that the only newborn infant in the troop at this time belonged to the handicapped female Guaca who almost never resisted attempts by caretakers to take her infant.

attempts) and Mopsa (who rebuffed in 19 of 155) foiled other females 12 percent of the time during their infants' first weeks of life. By contrast, Guaca, the relatively low-ranking, severely handicapped young mother in Bazaar troop was never seen to thwart a prospective caretaker in 59 attempts to take her newborn infant from her. Other factors besides ability to resist may enter into a mother's reluctance to give up her infant. These include age of the infant, the mother's parity (to some extent synonomous with experience), and her assessment of her infant's safety.

At this time, it is not possible to tease apart those interrelated variables such as youth, low rank, inexperience, and incompetence which increase the likelihood that the would-be caretaker will be rebuffed either by the mother or by another female. It is clear, however, that young, low-ranking nulliparas have the lowest success rate in taking infants. Based on data from the 1973-74 season, 68 percent of 196 abortive attempts were made by subadults, though overall subadults were responsible for only 54 percent of 747 tries.

In table 6.12, fifteen Toad Rock females are classified according to rank and experience: high-ranking experienced females who have had at least one infant; lower-ranking females who have had at least one infant; low-ranking nulliparas who are over fifteen months of age and who have had some experience in holding infants but who have never raised an infant of their own; and low-ranking nulliparas who are under fifteen months of age and who are relatively inexperienced with infants. An average rate of failure in attempts to take infants is calculated for each class, and between-class differences are tested statistically. From these calculations it is apparent that high-ranking experienced females have a significantly lower rate of failure than either older nulliparas or very young nulliparas. The two classes of older and younger nulliparas also have a significantly different failure rate from each other; very young females fail to get infants as much as 66 percent of the time. The failure rates of low-ranking experienced and older experienced females, and between high- and low-ranking experienced females were not significant. This lack of significance was due largely to the high degree of variability or sampling error within the lower ranking, experienced adult females. It should also be noted that the effects of low rank may be attenuated in some degree by the fact that during the 1973-74 season, subadult females were obtaining infants from other caretakers, usually other subadults, 78 percent of the time. Troop composition and the number of subadults, as well as the rank and reproductive status of the mothers of infants in the troop, may have important ramifications for the availability of infants to young females.

Egoistic Young Females and Their "Altruistic" Elders

Considering the near universality of hierarchical arrangements among social animals, surprisingly little is known about the factors that determine status. There are two known systems of rank ordering for mammals. In one, animals are ranked according to size or weight. In the second, genealogy, in conjunction with other developmental factors, determines rank. Anyone who has ever observed dairy cows in a field will not be surprised to learn that rank among cows is directly correlated with weight. In typical dom-

TABLE 6.12. Failure rates in obtaining infants for Toad Rock troop females, listed according to rank and experience.[a]

Category	Female	Failures	Mean and standard deviation[c]
A. High-ranking, experienced, adult females	Mole Mopsa I.E. T.T.	0.08 .13 .14 .14	Mean = .13 s.d. =0.25
B. Lower-ranking experienced, adult females	Scrapetail P-M C-E Pawlet	.57 .27 0 .17	Mean = .25 s.d. = .43
C. Lower-ranking, inexperienced, subadult females, over[b] 15 months	No. 4 Hauncha Handy Pandy	.33 .15 .24 .24	Mean = .24 s.d. = .064
D. Lower-ranking, inexperienced, females, under 15 months	Niza N-M Vert	.70 .71 .59	Mean = .66 s.d. = .055

a. Results of statistical tests:

Category	Result
A - B	n.s.
A - C	0.03
A - D	<.001
B - C	n.s.
B - D	n.s.
C - D	<.001

b. Only juvenile females were present in the troop when these data were collected.

c. Based on the total rebuffs in this class divided by total tries.

inance interaction, two cows approach, lower their heads, breath hard, stare at one another, and perhaps paw the ground. If neither animal retreats, physical contact ensues with one animal butting her forehead against the other. Intermittently, both partners draw back and each circles her opponent, maneuvering into a striking position against the unprotected side or flank of the opponent. A flanked female flees immediately. If flanking attempts fail, the two contestants again butt foreheads and each brings all of her weight to bear. Almost invariably, the heavier animal wins (Schein and Fohrman, 1955). In other ungulates such as mountain sheep, dominance bouts between males are decided by head-on charges; horn size is a reliable predictor of which male will win. If one male's horns are clearly larger than his opponent's, the bout may be decided without an actual contest (Geist, 1971). Apart from humans, rhesus and Japanese macaques provide the best-documented examples of nepotism. Daughters in these species fit into the hierarchy just below their mothers (Sade, 1967), and maternal rank also has some effect on male status (Koford, 1963).

Neither of these known systems of dominance can explain hierarchical relations among langur females. In the langur systems, smaller females are able to displace larger, heavier animals. Though from current information one cannot entirely rule out the effects of genealogy, clearly females at Abu are not inheriting a fixed position. According to the macaque literature, a high- or low-ranking female retains that same position throughout adulthood. A langur's rank, however, fluctuates dramatically throughout her lifetime. Furthermore, since young females rise to the top of the hierarchy, above older animals, some langur females would inevitably rank higher than their mothers.

The picture that emerges for langurs depicts a system in which old females gradually sink in rank while young ones rise. The result is a hierarchy top-heavy with females in their prime, between the ages of about four and ten. With the exception of alpha males, females in their prime have greatest access to priorities, while very old and very young females have the least. The relatively disadvantageous position of old females appears to be due to their disinclination to compete within the troop rather than to an inability to fight because of feebleness or disability, since these old females participate vigorously in intertroop encounters and in defense of troop members against outsiders. Even the very young, cream-colored infants and juveniles, regardless of their low status, unhesitatingly antagonize other animals by darting in and out at provisioned feeding sites or by challenging the monopoly of a larger animal at some food source. These youngsters may feed for a moment, but almost invariably they are threatened away or trounced soundly by a more dominant animal. Old females, however—despite the fact that they may be larger than the animals displacing them—simply do not try.

Four distinctive features emerge from the langur dominance

system. (1) Females in their reproductive prime compete for and obtain preferential access to advantageous positions and scarce commodities. (2) Old females do not compete and are at an apparent disadvantage when foraging, even when the females they are competing with are substantially smaller. (3) To all appearances, the above inequalities are due to the greater competitiveness and aggressiveness of young animals rather than to the inability on the part of old animals to fight. (4) Despite their timidity within the troop, these same old females remain active in troop defense.

A possible explanation for these curious features of the langur system derives from the combination of inclusive fitness theory with the concept of reproductive value. According to this hypothesis, an old female at or nearing the end of her reproductive career has a low reproductive value. She may stand to gain more in terms of genetic representation in surviving generations by investing in her close relatives and their offspring than by pursuing an egoistic course. By contrast, females at or approaching their reproductive prime (that is, having a high reproductive value) stand to gain in fitness by out-competing their other female relatives and by selfishly leaving matters of troop defense to other animals. Such a situation could explain the otherwise inexplicable altruism of old females who repeatedly went to the rescue of an infant attacked by an infanticidal male with greater vigor than the infant's own mother. Rather than being the exploitation of one class of individuals (old females) by another (younger ones), or representing a sacrifice on the part of old females, each female is actually maximizing her reproductive success in line with the life stage at which she finds herself.

Though high rank would not be uniformly advantageous throughout a langur female's lifetime, among macaques, high rank *would* be uniformly advantageous to the extent that a matrilineage encompasses members of all age-grades; that the rank of all members of the lineage are intertwined (as in Koford, 1963; Kawai, 1965a; Sade, 1972); and that inclusive fitness is cumulative for all members (Drickamer, 1974). The conditional phrasing here is dictated by the fact that some old female macaques *do* opt out of competition with age, though I have never seen an explicit reference to this in the literature. For example, in Iwamoto's food intake study, described earlier, he noted that two females who rated "zero" grain were so "old and timid" that they could not approach the feeding grounds, even though other members of their (generally low-ranking) lineage did feed (1974: figs. 1 and 7). Similarly, Kawai (1965a) mentions two examples of mothers who fell beneath their offspring in rank as they grew older. One possible explanation for these exceptions is that the arguments offered here also apply to some macaques. There is no reason to suppose that all macaques pursue the same strategies. Clearly, macaques in high-ranking lineages maximize fitness by close intralineage cooperation with relatives, but the advantages of cooperation might be devalued for members of low-ranking lineages; optimizing strategies for these animals might include more individualistic behavior in which rank is not independent of life historical patterns.

Acceptance of the hypothesis that langurs are ranked according to reproductive value must await proof that langur females do decrease in productivity with time. To date the case rests on only a handful of old females, Quebrado, Short, C.E, and Sol. Pending concrete information on age differences in the reproductive value of langurs, can a well-documented case be made for any other mammal?

Not long after I became convinced that rank among langurs depended on a female's reproductive value, I serendipitously encountered an article on dairy cows which pointed out that up until the age of about nine years dairy cows gain in dominance, and at around ten occupy the best social ranks in the herd. After the age of ten, however, animals show progressively declining dominance values (Reinhardt and Reinhardt, 1975). Schein and Fohrman's finding that heavier dairy cows displaced lighter ones remained true, because parallel to the age-dependent rise and decline in social rank, animals were gaining and losing weight. The Reinhardts report that cows gain weight up to the age of about nine years, when progressive weight loss sets in.

Immediately, there sprang to mind the prediction that, as hypothesized for langurs, dairy cows must decrease in fertility with age. A morning spent in the Boston Public Library with back issues of the *Journal of Dairy Science* paid off with a small treasure-trove of references, all pointing to the same phenomenon: with advancing age, dairy cows and range cattle decrease in fertility.[6] The domes-

---

6. Using the number of artificial inseminations necessary for conception as a measure of fertility among range cattle, Lawley and Bogart of the Missouri Agricultural Station (1943) found that fertility in heifers between two and three years of age was lower than among more mature animals. Only 66.1 percent of these youngsters calved, and they required 2.37 inseminations per calf. Fertility was highest in five- and six-year-old cows as shown by a calf crop of 86.2 percent and the requirement of 1.36 inseminations per calf. Beginning in the sixth year of age fertility gradually declined with increasing age until the ninth and tenth years, when the cows produced only a 69.2 percent calf crop, requiring 2.09 inseminations per calf.

tic importance of cows make possible sample sizes and a quality of data undreamed of by primatologists: "Since the beginning of the New York Artificial Breeder's Cooperative in June, 1940, until June 30, 1944, a total of 12,621 complete, recorded services to registered Holstein-Friesian cows, involving 41 bulls, have accumulated . . . The average number of services required per conception when based on all females, infertile cows included, was 2.07 . . . The influence of the age of the cow on breeding efficiency reveals a steady increase in the conception rate up to four years of age. Between the ages of five to seven years, inclusive, cows maintain a uniformly high breeding efficiency, which gradually declines with advancing age" (Tanabe and Salisbury, 1946). Decrease in fertility coincides almost precisely with decrease in weight and a consequent fall in rank. Whereas in langurs, old females would be "voluntarily" opting out of competition, in cows the matter is determined for them; animals forfeit high rank as their weight declines.

Both size-determined and nepotistic ranking systems rely on the capacity of the dominant animal, either by virtue of greater size or more powerful allies, to defeat subordinates if challenged. In the third type of ranking system hypothesized here for langurs, animals are ranked approximately in accordance with their reproductive value. Such a system would only be expected to occur among groups composed of close relatives, and depends to a large extent on the compliance of low-ranking animals. If the explanations offered are correct, both dominant and subordinate animals benefit from the inequality, making it very difficult in the case of female langurs to distinguish their dominance relations from cooperation.

7 / The Puzzle of Langur Infant-Sharing

This fable of an Ape, whiche had two children of the which [she] hated the one, and loued the other, which [she] took in her armes, and with hym fled before the dogges. And whanne the other sawe that his moder lefte hym behynde, he ranne and lepte on her back. And by cause that the lytyl ape whiche the she ape held in her armes empeched her to flee, she lete hit falle to the erthe. And the other whiche the moder hated held fast and was saued.

Aesop, ca. 620 B.C.
(tale translated from the Greek by
William Caxton, 1483; drawing
from a twelfth-century
English bestiary)

Maternal behavior is both widespread and relatively unsurprising, since by caring for her offspring, the mother invests in her own genetic representation in subsequent generations. More perplexing is the care of infants by females other than the mother. In the spirit of the British "auntie," meaning a close female friend of the family, such caretakers have become known in the primate literature as "aunts"—a term that does not imply relatedness to the infant and also does not preclude it. To circumvent possible confusion, the term "allomother" (or as it applies to both sexes, "alloparent") has been proposed to designate animals other than the mother who attend infants (Wilson, 1975). Whether we call them aunts or allomothers, the question remains: why do females take and carry some other female's infant?

The "Learning-to-Mother" Hypothesis

Allomothering was first described quantitatively for caged rhesus macaques at Madingley, England. Researchers there discovered that females around two years old were most eager to hold infants (Rowell, Hinde, and Spencer-Booth, 1964). More recently, Jane Lancaster (1971) has shown that among wild vervet monkeys in Zambia, a relatively small number of nulliparous young females were responsible for a disproportionately great number of contacts with infants. As Lancaster and others point out, most primates breed relatively late in life, give birth to only one offspring at a time, have long gestation periods, and, in some species, breed at only one particular time of the year. Hence the loss of an infant through negligence or inexperience will be costly. To the extent that maternal behavior is a skill that females may learn through practice prior to motherhood, allomothering benefits nulliparas.

The hypothesis that allomothering is practice for motherhood finds support among a wide variety of primate species in which females who have never had an infant of their own are especially eager to inspect and hold newborns. This nulliparous fascination with newborns has been reported for species in which allomaternal caretaking is common (vervets; bonnet macaques, Rahaman and Parthasarathy, 1962; gelada baboons, Dunbar and Dunbar, 1974a; squirrel monkeys, DuMond, 1968; howler monkeys, Glander, 1974; black-and-white colobus, Leskes and Acheson, 1971; Nilgiri langurs, Poirier, 1970; and hanuman langurs, Sugiyama, 1965a; Blaffer Hrdy, 1976); it has also been reported for species in which allomothering in the first weeks after birth is quite rare (among rhesus and Japanese macaques and savanna baboons).

Further support for the learning-to-mother hypothesis derives from the growing evidence that prior exposure to infants increases the likelihood of appropriate maternal responses. The literature on primiparous chimpanzee and monkey mothers has been reviewed by Lehrman (1961), who concluded that primiparas tend to provide offspring with less adequate care than multiparas do. In reconsidering these same observations, however, Seay (1966) found them inconclusive. In an experimental comparison of primiparous and multiparous wild-raised rhesus macaques, Seay found striking similarities in such maternal behaviors as cradling, restraining, retrieving, and nipple contact. The only significant difference he could find involved maternal confidence as reflected by the higher anxiety of the primipara, on the one hand, and by the higher percentage of physical rejections on the part of multiparous females and the increased firmness with which these experienced mothers rejected their infants. Seay concluded that "primiparous mothers normally give adequate care to their infants." But recent data from the rhesus macaque population on La Parguera Island tip the scale toward Lehrman's earlier conclusions. Drickamer (1974) found that only 50 percent of infants born to primiparous mothers survived to the age of six months. In contrast to the high mortality of firstborns, 90 percent of all rhesus babies born fourth or later in a female's sequence of infants survived to at least one year.

For caution's sake, it should be kept in mind that rhesus primiparas may be an extreme case. Rhesus mothers are unusually possessive of their newborns, and first transfer in this species occurs late. Compared to langurs, vervets, or gelada baboons, for example, rhesus nulliparas have little access to newborns prior to giving birth themselves. In fact, for just this reason good data on mortality by birth order among langurs might provide an interesting test of the learning-to-mother hypothesis. If the hypothesis holds, one would expect that the offspring of well-practiced langur primiparas would have a higher rate of survivorship than do firstborn macaques. There is nothing in my small sample of langur firstborns to contradict this prediction. At Abu, two langur primiparas, Scrapetail and Elfin, were not noticeably worse than other mothers, though in 1973 Scrapetail was—for a langur—exceedingly possessive of her infant and avoided females who might otherwise have attempted to take it. A third primipara, Guaca, did experience difficulty carrying her infant, but she was severely handicapped by the loss of her arm. Another primipara was briefly observed in the School troop during December of 1973. Like Scrapetail, this mother was extremely possessive of her newborn and refused to give it up to other females both before and after her infant was forcibly taken from her by a subadult from another troop. Counting this female, two of four primiparas were exceptionally possessive of newborns.

Whereas some primiparas are nonchalant or less competent, others provide their offspring perfectly adequate care. In some cases, individual variation and life history may be as important as parity. With these qualifications in mind, the La Parguera data suggest that when nulliparas do not have access to newborns, firstborn infants may be at a disadvantage.

These findings are supported by the results of cage studies. Rhesus macaque females who had been raised in isolation became abusive, even murderous, mothers. However, the same motherless mothers who were abusive with their first infant might care for their second and third offspring. Of six rhesus mothers who were indifferent or abusive toward first infants, five had second infants that received "adequate" treatment (Harlow et al., 1966). Similarly, a caged female gorilla who killed her first infant cared for a second two years later (Schaller, 1963).

The Langur Case: Methods and Subjects

Among primates, colobines permit allomothers to take newborns away from their mothers remarkably early. A hanuman langur may be taken from its mother within minutes or hours after birth and remain apart from her as much as 48 percent of the first day of life. Infants continue to attract attention from allomothers throughout the first months. Infant-sharing was described briefly by Charles McCann (1933b), but it was not until Phyllis Jay's work at Kaukori that there was careful documentation of the phenomenon (1965). Jay described how shortly after birth, just as soon as it was dry, the newborn langur is passed around among all the females in the troop.

Some intriguing questions remained unanswered, however. Exactly which females were attempting to take infants, and how often? Are they always successful? How do these females choose charges? And from whom are they actually obtaining the infant? How do the mothers and allomothers react to other females taking an infant from them? How long does each allomother keep her charge? How do different females treat their charges? How does the infant respond to these attentions? From the answers to these questions, it might be possible to learn why infant-sharing occurs.

In order to obtain quantitative answers to the above questions, I kept a careful record each time I saw an animal attempt to take an infant. A coding system was used to specify: the identity of the infant and of the allomother; the source of the infant; the reaction of each of the parties concerned (and in particular whether any of the three animals resisted the transfer); the mode of carriage of the infant once it was successfully taken away; the quality of the treatment dealt the infant; how the episode terminated (for example, the mother retrieves the infant; the allomother returns it to the mother; another allomother takes it; the infant is abandoned; and so on); and the duration of each allomaternal episode (Blaffer Hrdy, 1975: tables 4.22-4.24). Two different sampling techniques were used. In addition to the free-flowing ad libitum records that I normally kept of events as I noticed them, I also used focal-infant sampling to elucidate specific problems such as the success rate of different classes of females taking infants or the proportion of time that a newborn infant was with allomothers. In focal-animal sampling, as defined by Jeanne Altmann (1974), all specified interactions between a given individual and other group members are recorded during each sample period. For each sample, the length of time that a specified or focal infant was actually in sight was recorded. Such a sample period might range from several minutes in an extreme case to hours in another.

In 1973, only 39 percent of observed attempts to take infants were recorded during focal-infant sampling for periods totaling 1,153 minutes (19 hours). During the 1973-74 study season, periods of focal-infant sampling totaled 4,921 minutes (82 hours) and accounted for 76 percent of all observed attempts to take infants (table 7.1).

The purpose of focal-infant sampling is to minimize biases in data collection. Nevertheless, observations of semiarboreal, wild primates must be catch-as-catch-can. Although in some instances focal-infant samples were begun at a previously specified time (the ideal), in most instances, sampling was initiated opportunistically when the troop was undisturbed by either dogs, humans, or other langur troops, and when the targeted infant was clearly in view. Once begun, the vigil lasted until either a fixed time was reached or the infant was carried out of sight for a period of three minutes or more, whereupon the vigil terminated automatically. Hence, in many instances, the focal-infant sample depended on the visibility of the infant being observed. I do not believe this bias affected answers to major questions being asked, with certain important exceptions. Because subadult caretakers were more cautious of their charges, subadults were more likely to disappear from my

TABLE 7.1. Attempts to take infants divided into ad libitum and focal-infant observations.

Study season	Type of sampling	Number of observed attempts	Total duration of focal-infant samples (minutes)
February– March, 1973	Ad libitum	149	—
	Focal-infant	97	1,153
December– January, 1973-74	Ad libitum	144	—
	Focal-infant	459	4,921
	Total sample:	849	

occasionally "threatening" view. A statistical comparison of ad libitum records from 1973 and focal-infant records did not show any significant difference in the proportion of subadults represented. However, the possibility of a general underrepresentation of sub-adults in both ad libitum and focal-infant samples cannot be ruled out. A second and perhaps more important bias did lead to significant differences between the types of sampling. That is, the proportion of successful attempts to take infants was much higher in the ad libitum sample where 85 percent of all recorded episodes were successful ones. In contrast, only 71 percent of attempts to take infants recorded during focal-infant samples were successful. This difference could have occurred by chance only 2 times out of 100 and was probably no accident. Whereas abortive attempts to take infants are usually fleeting interactions, actually carrying the infant is a highly conspicuous and time-consuming enterprise. Unless the infant was already under observation when the abortive attempt occurred, there would be a bias against observing failures. Because of this bias, absolute rates of success can only be inferred from focal-infant samples. Success rates of individuals are probably unaffected and the entire sample, including both focal-infant and ad libitum records, may be used.

My investigation of allomothering focused on two troops, Toad Rock and Bazaar; a few observations were also made in I.P.S. and School troops as checks against idiosyncratic patterns in the focal troops. During the two seasons that data on allomothering were collected, a total of 18 infants were in Toad Rock and Bazaar troops, though the bulk of information was gathered for just 3 of them, and 68 possible caretakers were present (table 7.2). Since individuals are treated independently for each study season, the same animal may appear twice, once for each season. Older infants may be counted even more often, since any infant over about 13 months old, or else any infant that attempted to take another infant, would be listed both as a potential caretaker and as an available infant.

Among a number of primate species the youngest infants in the troop appear to be the most attractive to female conspecifics. For this reason, I decided that newborn infants would be the most fruitful targets for my observations. Three infants Brujo, Moli and Guat, were focused upon for periods totaling 82 hours, beginning shortly

---

1. A detailed treatment of which type of sampling would be appropriate for the analysis of various topics in allomaternal caretaking is provided in Blaffer Hrdy, 1975: table 4.G.

TABLE 7.2. The main study population for the focal-infant and ad libitum samples.

Period	Troop	Possible caretakers (including males)	Infants in the troop (sex and age)[a]
February and March 1973	Toad Rock	22	Niza (female, 1 month) N-M (female, 1 month) Brumio (male, ca. 11 months)
	Bazaar	13	Nuka (female, born in March) Astra (female, ca. 2 months) Quilt (male, 4-5 months) Bump (female, 4-5 months)
December and January 1973-74	Toad Rock	15 (or 23[b])	Brujo (male, born Jan. 9, '74) Moli (male, born Jan. 14, '74) N-M (female, 12-13 months) M-P (female, ca. 6 months) Niza (female, 12-13 months)
	Bazaar	18	Guat (female, born in Jan.) Elf (male, 3-4 months) Nonet (female, 10 months) Astra (female, 13 months) Bump (female, died Dec. 20) Infant kidnapped from School troop[c]
		Total: 68	18

a. Age is recorded for the beginning of the study period.

b. The number of possible caretakers was variable because of a political shift in which a new alpha male was able to oust nine resident males from the Toad Rock troop; three mothers and their offspring alternated between the ousted band and the newly reconstituted troop. Because males are so rarely involved in infant care and because the mothers returned to the troop during the observation period, the composition of caretakers in the newly reconstituted troop, including the wandering mothers (15), is used for the analysis.

c. The kidnapped infant was only present in Bazaar troop for one evening and the beginning of the following morning.

after the infants were born and continuing into the succeeding weeks. The number of days on which each infant was observed and the total number of minutes are summarized in table 7.3. Eighty-eight percent of 603 observed attempts to take infants during the 1973-74 study period involved one of these three infants.

In most of the recorded attempts to take infants, the identity of the

TABLE 7.3. Amount of time of focal-infant sampling between January 9 and January 31, 1974.

Troop	Infant	Days infant used as subject	Total minutes [hours]
Toad Rock	Brujo	23	2,587 [43]
	Moli	19	1,980 [33]
Bazaar	Guat	4	354 [6]

would-be allomother was known and it was possible to assign her to one of two categories: experienced or inexperienced. The term "experienced" applies to any adult female who has given birth to at least one infant. "Inexperienced" refers to a male or to a female who has never given birth to an infant. Only five of the cases (0.6 percent) involved males, and all of these males were immatures.

Inexperienced females composed only 26 percent of the female population but accounted for 54 percent of attempts to take infants. Within the category of inexperienced females, there was a striking difference in the competence exhibited by younger and older nulliparas. At Abu, nulliparas older than about one year were skilled caretakers, adept at climbing and running with an infant clinging to their chests. Should the infant fail to cling, an older nullipara typically pauses, reaches down with one hand, and boosts the infant up against her chest. If the infant continues to dangle, the allomother continues on her way, three-legged, one hand holding the infant against her body (fig. 7.1). In contrast, very young nulliparas (less than a year or so of age) frequently dropped, dragged, or deserted infants. As shown in table 6.12, very young females failed in their efforts to obtain infants significantly more often (66 percent of the time) than did older nulliparas (24 percent failures). Once an older nullipara obtained an infant, she kept it on average five times as long (about ten minutes) as a younger nullipara. Quite possibly, the greater persistence of older nulliparas in taking and keeping infants is due to their greater chance of being rewarded by success. Maturational differences might also contribute to this effect.

Several strands of evidence support the hypothesis that allomothering provides practice for mothering. Langur nulliparas display intense and prolonged interest in infants; older nulliparas are both more competent and more successful at taking infants than

Fig. 7.1. Subadult allomothers run bi- and tripedally, holding their infant charges against their bodies.

younger ones; prior experience with infants seems to be important in the normal development of maternal skills in a variety of species. Nevertheless, it is obvious that the learning-to-mother hypothesis is insufficient to explain *all* of the allomaternal behavior being recorded at Abu. That is, some 46 percent of the caretaking is undertaken by females who have had one or more infants, and remains to be explained. Kinship of course is one possibility, if by babysitting for close kin a female increases her inclusive fitness. Nevertheless, what little data there are on this question for Abu suggest that age of the infant and its availability are more important than degree of relatedness in determining which infant a female takes.

The most plausible answer to the question of why experienced females caretake comes from dividing all potential caretakers into categories based on their age and reproductive status. In 1973, the five categories used for analyzing the data were: nulliparous females; immature males; females who have recently given birth; pregnant females approaching term; and experienced females who are in the process of weaning an infant or who are not currently associated with an infant. In the 1973-74 study season, two new categories were added by subdividing old ones. Nulliparous females were divided into very young nulliparas less than about fifteen months of age, and older nulliparas from the age of approximately fifteen months until the last few months before the birth of their first infant. The second subdivision separates experienced females who are in the process of weaning an infant from experienced females who are not currently associated with an infant. Though, hypothetically, adult males are available as caretakers, at Abu they were never seen to take an infant except to attack it, and they are omitted from the following discussion.

The results of this division are provided in table 7.4. Three points stand out:

(1) In 1973-74 when older and younger nulliparas were separately considered, older animals were responsible for the greater proportion of attempts to take infants.

(2) In both the study seasons, experienced females weaning an infant or not associated with any infant attempted to take infants less frequently than would be expected on the basis of their representation among available caretakers. This lack of interest in newborns was particularly striking in some individuals. For example, on the day that the Toad Rock infant Brujo was born, Pawlet (who had a 13-month-old daughter at that time) passed within two meters of the newborn, which lay on the ground between its mother's legs, and paid no overt attention to it. In the 133 hours that Toad Rock troop was under observation during the 1973-74 study period, Pawlet attempted to inspect or hold an infant other than her own on only 6 occasions (compared to 84 occasions for T.T. or 49 for I.E., both females about the same age as Pawlet). The old female C.E. (who had no infant of her own at that time) attempted to take an infant on only 5 occasions. Based on data from both study seasons in the Bazaar troop, the two oldest females who were not then associated with infants exhibited almost no interest in caretaking. Likewise, Sol, the very old female in Hillside troop exhibited little interest in infants except to rescue them from infanticidal males.

(3) Taken by itself, point two—the lack of interest of some experienced females—is consistent with the learning-to-mother hypothesis. Nevertheless, data from the 1973-74 study season indicate that pregnant females may be overrepresented among would-be caretakers, regardless of whether they have had previous maternal experience. In 1973-74, four pregnant, experienced females (Mopsa, Mole, T.T., and I.E.) who constituted 12 percent of the available caretakers were responsible for some 26 percent of all caretaking attempts registered. Due to the comparatively few observation hours spent with the Bazaar troop, only a very small portion of all caretaking was attributable to the primiparous pregnant females Guaca and Breva. Hence, the overall representation of pregnant

TABLE 7.4. Allomaternal attempts, including both successful and abortive tries, divided according to reproductive status of the allomother for 1973 and 1973-74.

Number and proportion	Nullipara		Immature male	Lactating mother	Pregnant female near term	Experienced female with weaning or no infant		Total
1973								
Number of attempts	124		4	17	11	28		184
Proportion of total attempts	.67		.02	.09	.06	.15		≈100
Proportion of available caretakers	.21		.24	.18	.03	.33		
1973-74	Under 15 months	Over 15 months				Weaning	No infant	
Number of attempts	84	209	1	97	158	28	19	596[b]
Proportion of total attempts	.14	.35	.002	.16	.26	.05	.03	≈100
Proportion of available caretakers	.19	.16	.06	.16[a]	.13[a]	.16	.13	

a. The number of females present in these categories fluctuated as pregnant females gave birth during the study period.
b. Total of 780 represents all caretaking attempts in which the reproductive status of the allomother was known, regardless of whether she was individually recognized.

females among interested caretakers could not have been inflated by the presence of these two inexperienced females.

In addition to the overrepresentation of pregnant females among caretakers, lactating females (16 percent of the available caretakers) were also well represented during the 1973-74 study period and accounted for 16 percent of caretaking attempts. Compared to females who were not currently associated with infants, these females with recently born infants displayed greater interest in infants. The high percentage of caretaking attributable to recent mothers was due largely to Mopsa and Mole, two females who gave birth during the study period and who repeatedly held each other's infants, which were five days apart in age.

Because few animals were in each category, the possibility of variation between individuals is troublesome. In order to increase the number in each category, data from two seasons were pooled. New proportions for the combined data were calculated even though sampling techniques were not equivalent: a larger proportion of the 1973 observations were ad libitum, while most made in the 1973-74 study season were during focal-infant samples. These combined data are presented in table 7.5. Chi squares calculated for the overall differences between the proportion of caretaking attempts observed versus those expected on the basis of each category's representation within the population of available caretakers were highly significant. It is unlikely that chance alone accounted for the observed variance in caretaking attempts.

One way of describing the high frequency with which some females attempt to take infants is to refer to them as "motivated" to hold and carry infants. Not surprisingly, there was a correlation between those females most motivated to take an infant, and those seen holding more than one infant at once. In 16 of 26 such cases, the "super caretaker" was a recent mother, and one of the two infants being held was her own. Similarly, on one occasion a weaning mother held her own older infant briefly in conjunction with a younger one. Despite the awkwardness involved, childless females sometimes obtained and held two infants. An older nullipara held two infants on two occasions, and pregnant females on seven.

At Abu, motivation to take and hold infants is far from uniform. Among experienced females without infants or with infants that are being weaned, a newborn does not seem to be a "focal point of interest" for *all* langur troop members (Jay, 1963b). Nor is it necessarily "the female without a dependent infant of her own" who spends time borrowing the small infants of other females (Ripley, 1965;

TABLE 7.5. Summary of allomaternal attempts for the 1973 and 1973-74 study periods.

Number and proportion	Nullipara	Immature male[a]	Lactating mother	Pregnant female near term	Experienced female with weaning infant or no infant	Total
Number of attempts	417	5	114	169	75	780
Proportion of total attempts	.53	.006	.15	.22	.10	≈100
Number of caretakers	18	10	8-11[b]	5-8[b]	21	65
Proportion of available caretakers	.28	.51	.12-.17	.08-.12	.32	≈100

a. Even if immature males are excluded from the sample, nulliparous and pregnant females are still significantly over-represented. Excluding males: $X^2 = 338$; d.f. = 3; $p < .001$.

b. Number fluctuated because some females gave birth in the course of the study period.

Sugiyama, 1965a), since lactating females may be heavily involved in infant-sharing. More precise statements concerning the reproductive status of the females involved are needed. The next question, of course, is why should reproductive status matter?

Accumulating evidence from a variety of species lends support to the hypothesis that some allomothers are in the process of learning skills. Beyond providing naive females with vital mothering practice, prior exposure to infants may lower the thresholds at which both naive and practiced females respond maternally to infants. Such an increase in responsiveness is referred to in the psychological literature as "sensitization" or "priming" (Hinde, 1970). In reviewing the onset of maternal care among rodents, Noirot (1972) concludes that "exposure to pups, provided it lasts long enough, invariably leads to the development of maternal behavior in naive rats, mice and probably hamsters." Repeated exposure to pups is sufficient to elicit maternal behavior in immature and adult rats, and the sensitization period is of comparable length in adult intact males, virgin females, and castrated males and females. There is as yet no conclusive evidence that analogous sensitization operates among primates.

If, indeed, some form of priming does occur among monkeys, exposure to infants would be most adaptive for nulliparas and pregnant females, who must respond to their infants positively and with little delay at the moment of birth. Mothers of highly dependent, suckling infants might also profit from a lowered threshold of responsiveness to infants generally. Although in some cases the infant eliciting a response might belong to some other female, the suckling mother's chances of promptly responding to her own infant might also be increased. Since a langur mother very soon recognizes her own infant, she is safeguarded from investing substantially in any other infant. At Abu (with one possible exception), no mother was ever seen to allow any infant but her own to suckle. Furthermore, caretaking by lactating females is fairly short in duration, an average episode lasting 3.93 minutes. One such mother (Mopsa) was observed simply to desert a borrowed infant on three different occasions after taking it and then apparently tiring of it.

Priming can only provide part of the explanation for caretaking by pregnant females. What priming can not explain is why females initiate contact with infants in the first place. Although endocrine changes are not essential for establishing maternal behavior in rodents or primates, the problem of initial contact and recurrent contacts can be resolved by postulating endocrine changes that increase responsiveness to infants during gestation. Support for this view is provided by Terkel and Rosenblatt's (1968) discovery that the onset of maternal behavior in rats could be accelerated by injecting blood taken from females who had just given birth. These findings are also consistent with the gestation effect noticeable just after midpregnancy among rats and mice in several measures of maternal behavior (Noirot, 1972). Comparable data for primates, however, do not exist.[2] If there is a hormonally induced rise in responsiveness to infants during pregnancy, the continued but lessened interest in caretaking exhibited by lactating mothers among the langurs at Abu might well represent the tailing off of this pregnancy phenomenon. Pending further information, both priming and endocrine changes during pregnancy are tentatively postulated here. Whether caretaking by pregnant females is a behavioral pattern that has actually been selected for or whether it is simply a "beneficial consequence" of hormonal changes that evolved for quite different reasons (Hinde, 1975) remains to be seen.

Information Gathering

If the advantages hypothesized are real, practice and priming together could account for most allomaternal episodes observed at Abu. Ten percent of caretaking events, however, involved experienced mothers that were neither pregnant nor lactating. Barring the possibility that such caretaking is altruistic, these females' interest in infants remains unexplained. One possibility is that all females gain information from rudimentary examination of an infant who is both a fellow troop member and a relative. By picking up an infant once, or a few times, allomothers can quickly determine its sex and condition.

The allomaternal practice of holding an infant upside down and sniffing, peering at, and inspecting its genital region is found among a variety of primates, including baboons, vervets and langurs (fig. 7.2). It is difficult to imagine an explanation for this practice other than information gathering. Curiously, although the degree of motivation to hold an infant and the length of time she will keep an infant can be fairly well predicted by a female's current reproductive status, the same is not really true for genital inspections. Experienced females were more likely than inexperienced females to per-

---

2. Although a "gestation effect" has never been specifically reported for a primate species, it is interesting that squirrel monkeys exhibit a pattern similar to that of langurs; that is, nulliparas and pregnant females show the greatest interest in infants. In his study of caged squirrel monkeys Rosenblum (1972) found that pregnant females near term were the most likely to retrieve infants.

Fig. 7.2. A multiparous allomother holds a newborn infant upside down by its tail in order to inspect its bottom.

form genital inspections and did so more or less in accordance with how often they had the opportunity to do so.[3]

Based on these findings, it seems plausible that all females, but perhaps especially experienced females, are interested in a summary investigation of the new infant. Accordingly, we would expect each female troop member to take a newborn infant once or a few times during its first weeks of life. Only some—nulliparous, pregnant, or recently delivered—females will take it again and again.

Costs and Potential Benefits to the Infant

In all probability, allomothering benefits the caretaker. Whether or not an allomother will be able to obtain an infant, however, ultimately depends on the reaction of the infant, and especially the infant's mother, who to a large extent controls access to her offspring. Their view of the matter must also be considered.

As with most other primates, langurs tend to give birth at night or in the early morning hours (Jolly, 1972b). Because of the problems this timing creates for field observers, all records from the wild begin after the infant langur is several hours old. At the San Diego Zoo, however, James McKenna (1974a, b) was able to observe the daytime birth of a male langur infant. According to his report, juvenile females continually fought for a position near the emerging infant. In the seconds following the birth the mother denied group members access to her infant, but not more than 15 minutes after parturition a juvenile female gently took the newborn infant and held it without any distress being exhibited by the mother.

Newborn langurs enter the world totally dependent on their mothers for nourishment and on mothers or allomothers for mobility and security. The infant's greatest contribution is to cling to the hair of whichever female has it. In some instances—for example, when the mother stands up abruptly or when she rears on her hindlegs to feed on an overhanging food source—the infant cannot catch hold properly and the mother must hold the infant with one hand. Often the infant fails to orient properly and must be adjusted so that it clings ventro-ventrally upright on the chest when the caretaker is

---

3. The multiparous female T.T., who took infants more often than any other adult female, also turned infants upside down more often (on 19 occasions). Other experienced females who turned infants upside down included: Mole, who did so on 15 instances; I.E., who did so on 11; P-M and Mopsa, each 4 times; Pout, 3 times; Overcast, 2; and Nony, Kasturbia, Earthmoll, and Pawlet, once each.

seated, or head forward if she is walking. One consequence of such extreme neonatal dependence is that it may prove unwieldy for either a mother or an allomother to feed while carrying an infant.

Although Sugiyama (1965a) suggests that eyesight does not develop until the fifth day, at Abu infants have their eyes open and appear to grope toward objects in their environment as early as the first day. Even if the infant could discriminate objects in its environs, body coordination is poor during the first hours. When disoriented, infants typically flail their arms about, especially if they lose the nipple. This flailing is adaptive insofar as it leads the infant to brush against the appropriate furry surface or to relocate the nipple. If an infant is radically dislocated, and not adjusted by the mother or allomother, the infant's strategy is to cling however it can to the nearest animal's rump, arm, or back.

By the end of the first week or so, the infant orients itself and holds on expertly, quite able to handle itself while its mother forages. Despite increasing competence, the infant has little control over which animals hold it during the first weeks of life. Unless the mother herself resists transfer by moving away, huddling over her infant with her body, or threatening the would-be caretaker (fig. 6.7), allomothers may forcibly pull an infant from its mother. Once taken, the infant invariably clings to the new caretaker.

Infants do not exhibit any overt preference for their mother until around the second week of life. Nevertheless, almost from birth infants appear to be aware of the distinction between their own mother and other females. An infant that has been quietly holding its mother may begin to whine soon after being taken by some other female. Conversely, an infant that has been struggling and complaining will usually grow quiet when retrieved by the mother. It is unlikely that this discrimination depends on maternal competence, since the transformation in infant behavior may occur regardless of how a mother holds her infant. The three-legged mother Guaca, for example, experienced difficulty adjusting her infant, Guat, while moving; consequently, Guat was often attached to her mother in various semiprecarious perches, similar to the positions taken by infants clinging to unsolicitous allomothers. Nevertheless, Guat appeared to prefer her mother to a more coordinated allomother. Very probably, infants are discriminating—at least at the outset—on the basis of whether or not they are allowed to suckle.

By the fifteenth day (in the case of Brujo) or slightly earlier (in Moli's case), allomaternal caretaking begins to be complicated by the infant's predilections; if the infant resists transfer, the allomother may occasionally fail to take the infant even though the mother apparently accedes. Still, at all ages, the failure rate due to infantile reluctance is low (6 percent—but note that this figure is based primarily on data from infants less than two months old). At 16 days old, Brujo first reached out to his mother, who also reached out to him. In this way, Brujo first initiated transfer back to his mother. At 20 days old, Brujo was two meters from a subadult female (Hauncha) who had taken him; Brujo was able to crawl the distance back to Hauncha on his own accord. The next day, Brujo, an accomplished crawler, left his mother to approach a cluster of grooming females. By the third week, Brujo himself terminated 30 percent of allomaternal episodes by returning on his own to his mother. The infant's interest in visiting allomothers grows with the infant's ability to explore its environment. At the same time, the infant's attractiveness to allomothers is decreasing. Allomothers will rarely take and hold an infant several months old if a younger infant is available.

Although allomothers appear to prefer younger infants, the cost of allomothering to the infant is probably greatest for newborns, who may have difficulty feeding in the very first days after birth. On the first day of life, some infants scarcely suckle at all (although nothing is known about nighttime feeding). For example, Moli, the daughter of a young female whose nipples had not yet become pendulous with nursing, was unable to keep her mother's nipple in her mouth. On several occasions when Moli had, after much groping, successfully taken the nipple in her mouth, the mother raised up one arm to feed and the infant lost the nipple again. In addition to difficulty locating and retaining the nipple, a more important impediment to nursing was the continual removal of the infant from the mother by allomothers. Records kept for two Toad Rock infants, Brujo and Moli, indicate that in the first week Brujo (born January 9, 1974) was on his mother only 55 percent of daytime. Allomothers took the infant for the remaining 45 percent of 1,516 minutes that the infant was the subject of focal-infant sampling. Moli (born January 14, or the night of January 13), spent a smaller proportion of daytime (39 percent) off her mother during 1,192 minutes of observation in her first week. Quite probably this decrease in allomaternal interest in the second newborn was due to the fact that the same number of motivated allomothers could now divide their attentions between two comparably new and available infants (both mothers refused to give up the infants about 12 percent of the time).

With the birth of Moli, his younger competition for allomothers

Brujo spent more time on his mother (73 percent) during his second week of life than he had in his first. In contrast, Moli spent more time on her mother in the first week (61 percent) and slightly less in the second (50 percent). Over all three weeks, both infants spent roughly the same amount of time off their mothers. Of 1,544 minutes of observation in his first 21 days, Brujo was off his mother 40 percent of the time. In 1,980 minutes during her first 18 days, Moli was off 43 percent of the time.

By a fortunate coincidence, the only comparably quantitative data published on the time that a langur newborn spends off its mother in the first days of life come from a caged group of approximately the same size as the Toad Rock troop, though it contained fewer juvenile females and more juvenile males. This troop of one adult male, eight adult females, four juvenile females, four juvenile males, and two infants was studied in the San Diego Zoo (McKenna, 1974a). In 70 hours of observation, one newborn male infant was off his mother 53 percent of the time, not substantially more than the infants were separated from their mothers in the wild group.

The parallel finding in both wild and cage studies that langurs spend about one half of their time off their mothers suggests that prolonged separation from the mother is normal for this species. Since infants have never been weighed in any langur study, the effect of long separations from the mother on nutritional intake is not known. Weight gain among newborns has been a subject of inquiry among rodents, however. In their study of communal nursing among house mice, Saylor and Salmon (1971) found that by 19 days after birth the weights of pups belonging to mothers who had been housed with nonlactating virgin females tended to be lower than the weights of pups whose mothers were housed alone or with other mothers and their pups. In fact, when the second female present was lactating, pups tended to weigh more, even though the experimenters maintained the same ratio of lactating females to pups. One explanation for these findings is that the allomaternal behaviors (including nest building, retrieving, and hovering over the young) of the nonlactating virgins were interfering with pup suckling activities.

Comparable studies have not yet been undertaken for primates, but cases of interference by allomothers resulting in infant starvation are known. Among caged squirrel monkeys, the death of six infants as a result of maternal reluctance or inability to retrieve their infants from caretakers has been reported (Rosenblum, 1971). Among a wild group of West African Lowe's guenons, allomothers took a newborn infant from a temporarily ill mother. After the mother recovered, the caretakers refused to give up the infant and threatened the mother away whenever she approached. The infant died after four days, apparently from starvation (Bourlière et al., 1970). Among hanuman langurs, intertroop kidnappings have led to death in at least two instances (Mohnot, 1974); in other instances of near-starvation, either the observer intervened (Sugiyama, 1965a) or else the mother was eventually able to retrieve her infant.

Potential cost to an infant of temporary separation from the mother is unknown. Milk samples from baboons, macaques, orangs, chimps, gorillas, and man are very low in protein content (from 1.2 to 2.1 percent); fat content is also low (3 or 4 percent) for all species except baboons (3.4 to 19.4 percent) (Ben Shaul, 1962a, b). From such data Blurton Jones (1972) hypothesized that higher primates are continuous feeders, adapted for near-constant access to the nipples. If true, we would expect langur babies to suffer from long periods off the mother. Alternatively, colobine mothers may compensate by producing richer milk than that of other primates. Unable to obtain samples in this country, I asked Dvora Ben Shaul in Israel if she had analyzed any colobine milk. She replied that she had tested one sample for her earlier study, but protein and fat contents were so high (14 and 12.5 percent, respectively) that she assumed an error had been made. If her unpublished findings are substantiated by further analyses, they strongly support the hypothesis that colobine mothers compensate for infant-sharing by producing richer milk.

Although the life of an infant langur appears to be a harrowing one, injuries due to allomaternal negligence or brutality are rare under natural conditions. In thousands of hours that langurs have been observed at Dharwar, Orcha, Kaukori, Polonnaruwa, Jodhpur, and Abu, only on a few occasions were infants dropped or dragged so roughly that bleeding resulted (Mohnot, 1974; this study); most such injuries were in the form of small cuts in the head region. At the San Diego Zoo, however, where the floor surface is hard and rough, dragging of infants by allomothers has resulted in a number of visible injuries, including bloody noses and scraped and bloodied heads and tails (J. McKenna, personal communication). The evidence to date suggests that serious injury to infants from allomothers must be rare; it also seems unlikely that such caretaking is being performed altruistically and solely for the benefit of the infant (fig. 7.3).

Given the potential cost to an infant from allomothering, why does a mother give up her infant? It seems possible that a mother may be

Fig. 7.3. Guaca's infant is held by Earthmoll, a multiparous, middle-aged female who ignores the fact that Guat has slipped down between her legs; the whining expression on Guat's face is typical of a complaining infant.

taking out a form of insurance against the day when she is temporarily incapacitated or absent and danger threatens her infant. This could be so only if a former allomother is more likely to rescue an endangered infant. Yet the old langur females such as Sol and Quebrado who exhibited the greatest daring in defending infants very rarely exhibited interest in holding an infant. They were apparently available as rescuers even though they exhibited little interest in routine allomothering. There may also be socialization benefits to be derived from an infant's exposure to other troop members. More plausible, perhaps, is the hypothesis that a mother gains immediate benefits from the freedom to forage unhampered by a dependent newborn. It was my distinct impression that mothers or caretakers with newborn infants foraged very little while holding them. Unfortunately, the focal-infant sampling method was incompatible with watching what the mother did after she gave up her infant; quantitative data on how much the mother foraged with and without her infant were not collected. Presumably the advantages to the mother from freedom to forage decrease with time as the infant grows more coordinated and independent.

Other factors that influence a mother's compliance with allomothers may include her assessment of current security; her parity (for example, some primiparas are unusually possessive); and her assessment of the prospective caretaker's age, status, and competence. As shown in table 6.12, very young (under 15 months), low-ranking, and inexperienced females were more likely to be rebuffed by both mothers and allomothers. Maternal rank may also affect the availability of the infant; the single case supporting this view is that of the handicapped female Guaca. Low-ranking Guaca not only never rebuffed other females who tried to take her infant, but she herself was frequently rebuffed when she attempted to retrieve her infant—even from females younger and smaller than herself.

Why a mother retrieves her infant, or why she does so when she does, is not well understood. As can be seen in table 7.6, the age of her infant is not a helpful indicator of what proportion of caretaking attempts will be terminated by maternal retrieval. A newborn infant such as Brujo is not necessarily retrieved by his mother more frequently than an infant 3 months old such as Astra. Age, however, is a relatively good predictor of how attractive an infant will be to other allomothers. Infants less than one month old (Nuka, the I.P.S. infant, Niza, Moli, Brujo, and Guat) typically were taken from allomothers by other allomothers more than 50 percent of the time. Though at this time it is not possible to assign weights to the various factors influencing the mother to retrieve her infant, a list of them must include: her infant's distress signals, the actual distress of her infant, and satisfaction of the mother's foraging needs. It is worth noting, however, that a mother absorbed in foraging or grooming may ignore her infant's cries, while an allomother responds with greater alacrity. In five of ten cases where an infant was deserted, an allomother retrieved it.

Attractive, Available, but Not Necessarily Related Infants

The success of an allomother in acquiring an infant depends partly on her ability to take one from its mother and partly on her ability to take one from other allomothers. Data from 1973-74 show that allomothers obtained infants from their mothers only 30 percent of the time; more than 55 percent of the time, allomothers obtained infants from other caretakers. The Bazaar troop infant Nuka provides an example of an infant whose exceptionally possessive young mother, Kasturbia, rebuffed prospective caretakers 20 percent of the times they tried. In Nuka's case, allomothers took the infant directly from the mother in only 13 percent of their attempts. In contrast, Astra, whose multiparous mother, Pout, was more permissive and refused caretakers access to her infant only 3 percent of the times they tried, was usually (52 percent of times) taken directly from her mother.

TABLE 7.6. Termination of 561 allomaternal episodes, based on observations of eight infants present during the 1973 and 1973-74 study seasons in the Toad Rock, Bazaar, and I.P.S. troops.

Infant	Mother retrieves her infant (%)	Infant returns of own accord (%)	Allomother returns infant to mother (%)	Another allomother takes infant (%)	Allomother abandons infant (%)	Number of terminations witnessed[a]
			Mode of termination			
Astra (ca. 3 month)	38	33	0	29	0	21
Nuka (ca. 1 month)	12	30	0	52	6	33
I.P.S.[b] (newborn)	40	7	0	40	13	15
Niza (ca. 1 month)	23	16	0	53	6	64
N-M (ca. 1 month)	33	5	11	11	39	18
Moli (newborn)	40	0	1	53	5	173
Brujo (newborn)	34	8	0	57	0.5	191
Guat (newborn)	24	9	0	63	4	46

a. It is assumed that all five types of termination were equally likely to be observed.
b. This infant was only observed during one day; the data are included here for the purpose of increasing the number of cases for comparison.

Whether or not an allomother approaches the mother or another allomother may depend on the individual and the individual's likelihood of success, but some trends with age and reproductive categories were also apparent. All reproductive categories except recent mothers were more likely to take an infant from another allomother than from the real mother; recent mothers are exceptions because of the frequent exchange of infants between two high-ranking recent mothers, Mopsa and Mole. Nulliparas in particular most often took infants from other allomothers, doing so 72 percent of the time (Blaffer Hrdy, 1975).

Once an allomother has an infant, how long she keeps it depends both on her own choice and on the desire of another langur (either the mother or another allomother) to take it from her. An allomother who tires of her charge has the option of pushing the infant off or abandoning it. Usually the screams of an infant being pushed off or pressed against the ground will attract another allomother before the infant can actually be abandoned. Hence, abuse by the allomother functions (inadvertently?) as a means of inducing another female to take the infant from her. In some cases, however, other females take the infant from an allomother *before* she is willing to give it up, particularly in the case of subadult females competing among themselves for access to an infant. The high motivation of these nulliparas to keep infants, the relative difficulty for them of getting infants, and their vulnerability to competing caretakers may combine to make young females evasive in their caretaking patterns. Nulliparas could be distinguished from all other allomothers by their practice of taking the infant and running. This skittishness creates the sampling problem discussed above. To avoid interference from other females, nulliparas were far more prone to carry an infant into thick brush or up a tree and out of sight. Nulliparas also avoid human observers more than other females do, perhaps because people are a potential source of danger and their proximity might increase the likelihood that the mother will retrieve the infant.

As far as I can determine, the attractiveness of an infant to would-be caretakers is a composite of neonativity, availability, and the presence or absence of alternate choices (either newer infants or older infants that are more accessible). One possible measure of an infant's attractiveness is how many attempts per hour are made by allomothers to take it.[4]

---

4. If allomothers are conditioned by success, the number of overall attempts may also be correlated with past successful attempts, complicating the issue of "attractiveness."

Table 7.7 presents the rate of allomaternal attempts per hour made on each of eight infants under three months of age present in the I.P.S., Bazaar, and Toad Rock troops in either 1973 or 1974; a ninth infant born in the School troop but temporarily kidnapped by Bazaar troop females is also included. To the extent that rates of allomaternal attempts can be used as measures of attractiveness, two infants stood out for their popularity: Guat, the singleton new-born of a low-ranking, handicapped, and very permissive mother, and especially the infant kidnapped by Bazaar troop females. Allo-mothers attempted to take Guat on average 6.6 times per hour while she was the subject of focal-infant sampling; during one hour of focal-infant sampling, 13 attempts were made to take the kidnapped infant. In both cases, no other infant was present in the troop to compete for allomaternal attentions.

By contrast to these very available infants, Moli, the second of two infants born in the Toad Rock troop in January and the daughter of the mildly possessive alpha female, was approached less often (3.6 tries per hour), even though the Toad Rock troop contained more potential caretakers and more motivated caretakers than the Bazaar troop. Brujo, the first infant born in the Toad Rock troop, was aunted slightly more than Moli (5.4 tries per hour), even though Brujo's beta-ranked mother was equally possessive, possibly due to the effects of birth order on attractiveness.

A crucial question for studies of caretaking behavior is whether or not allomothers are more attracted by close kin. Since there are no long-term genealogical data for Abu, it is impossible at this time to be certain; nevertheless, four lines of available evidence may foreshadow the answer. These suggest that infants are not being chosen on the basis of either relatedness or familiarity (which under natural circumstances is usually concomitant with genetic related-ness). First, the most reliable predictors of how motivated a langur will be to caretake are sex, age, and reproductive state; motivation to caretake fluctuates with reproductive condition regardless of which infants are present and how closely they are related. Second, individual preferences for specific infants are rarely apparent. Rather, any infant of approximately the same age can serve as a substitute for any other infant, and it is not unusual to see an allo-mother switch her attention from one infant to another if the first in-fant proves inaccessible (Blaffer Hrdy, 1975). Third, in two cases I was fairly certain that infants had nulliparous siblings present in the troop. One of these older sisters (Handy), to the extent that she preferred any infant at all, preferred the infant Brujo, who was born five days earlier than Moli, her putative half-sibling. The other

TABLE 7.7. The attractiveness of nine infants under three months of age born in the Toad Rock, Bazaar, and I.P.S. troops during 1973 or 1974, as measured by the average number of allomaternal attempts per hour to take them.

Infant	Likelihood of maternal rebuff[a]	Other infants 3 months or less currently in troop	Approximate age of infant during observations	Approximate number of allomaternal attempts per hour
Astra	0.03	Nuka and Bump[b]	3rd month	5[c]
Nuka	.21	Astra and Bump	1st-2nd month	3.7
I.P.S. newborn	.06	Several	1st week	5[c]
N-M	.26	Niza	1st-2nd month	3.6[c]
Niza	.11	N-M	1st-2nd month	3.2
Brujo	.08	Moli born when Brujo 6 days old	1st three weeks	3.8
Moli	.07	Brujo	1st three weeks	3.5
Guat	0	None	1st week	6.6
Infant kidnapped on December 23 by Bazaar troop females	0	None[d]	1st week	13[c]

a. The willingness of a mother to give up her infant to another female is not the only component of infant availability; in particular, the likelihood of maternal retrieval and of interference by another allomother must be taken into account.

b. Observations of Bump totaled less than one hour and were too short to warrant her inclusion among the infants whose attractiveness is measured.

c. Based on very short time samples.

d. Kidnapping occurred in the month before Guat was born.

suspected sibling (Hauncha) did exhibit special interest in her suspected full-sister (Niza), but it was impossible to separate "preference" from the possibility of privileged access—that is, Hauncha was never rebuffed by her mother (Pawlet) when she attempted to take Niza. This could have been due either to preferential treatment of close kin or to the fact that Hauncha was an older subadult and a very competent caretaker who might have been allowed access to the infant in any event. In four additional cases in which there was a slight preference exhibited for a specific infant, I was certain that none of these shared a mother in common, and they probably did not share the same father.

The final and perhaps most convincing line of evidence is the great interest manifested by langur females in infants born in other troops, infants which almost certainly are less closely related than infants born in a female's own troop would typically be. Whether or not alien infants are taken seems to be purely a matter of availability. Because of the antagonism between females belonging to different troops, infants may be seen and heard by females in neighboring troops but are rarely accessible to them. Nevertheless, females do occasionally succeed in kidnapping infants from neighboring troops and in keeping them for extended periods of time. When such kidnappings occur, they provide us with a natural experiment to show how an unrelated infant without a mother to protect it is treated by females.[5] As illustrated in the following example (continued from the kidnapping episode described in chapter 6), kidnapped infants are very attractive to allomothers but perhaps less well-treated than more closely related infant troop members whose mothers are present. In the following account each change in possessor, or attempted transfer of the infant is numbered.

December 23, 1973 (continued). The mother from School troop attempts to retrieve her infant, but the newborn is retained by Bazaar troop females. (1) At five p.m., subadult Junebug, carrying the kidnapped infant, joins other Bazaar troop females as they move toward their sleeping site at Sanand House. (2) Wolf takes the newborn from Junebug. The infant tries to suckle from Wolf (who recently lost her own infant) but is pulled down from the nipple by Wolf. (3) Pregnant Guaca approaches, nuzzles, and tries to take the infant, but Wolf moves away from her. (4) At 5:05, Guaca again tries to take the infant

5. The reaction of male langurs to the presence of a kidnapped infant in their troop is also interesting. Whereas males just taking possession of a harem kill unfamiliar infants, a male ignores an alien infant whose mother is not present.

from Wolf and succeeds. (5) Wolf retrieves the infant from Guaca. (6) Junebug then takes the infant from Wolf. The infant clings to Junebug as she makes a long leap between two trees. Later, Junebug pulls the infant off herself and turns it upside down. As the infant screams, Junebug turns her upside down again and bites her rump. Moments later, as Junebug moves again, the infant slips off her stomach and begins to scream. (7) Wolf takes the infant. As the infant continues to writhe and to scream, and also to attempt to get at Wolf's nipple, Wolf lifts up her rear foot like a dog scratching fleas and uses her foot to push the infant down to the ground. (8) Nonet, scarcely more than an infant herself (and who does not normally have access to infants), then takes the kidnapped infant from Wolf. Nonet holds the infant tightly to herself as she makes a somewhat precarious leap into another tree. (9) Elfin takes the infant from Nonet; she sits holding the stolen infant and her own older infant, Elf. Though I could not see Elfin's nipples, it is possible that the kidnapped infant did nurse because she grew silent. Moments later, Elfin stands to move. The stolen infant has slipped down onto Elfin's stomach and cannot cling properly. Elfin kicks the squealing infant away, then pauses to adjust both infants to her. To do so, Elfin sits down again. (10) Junebug takes the infant from Elfin. Almost immediately, Junebug uses her back foot to push the infant off her. Junebug feeds as the infant clings quietly to her. (11) At 5:45 Breva takes the infant from Junebug. The infant screams and tries to remain on Junebug. (12) Less than one minute later, an unidentified multipara attempts to take the infant from Breva, but Breva refuses to give up the infant. (13) At 5:47 Overcast takes the infant from Breva. The infant clings quietly to her chest until Overcast turns the infant upside down and grooms her bottom at which point the infant resumes screaming. Overcast pushes the infant off her and with both hands presses the infant face down against the limb on which they sit. With few respites, the kidnapped neonate has been crying for almost the full hour between five and six p.m., when observations terminated.

If, as appears here, kidnapped infants are poorly treated, the next question is why would a mother give up her infant to alien allomothers? From the description of a kidnapping at Dharwar provided by Sugiyama (1965a), it appears that the mother did not notice that the female who took her infant was an alien until it was too late. At Abu the mother may simply have proved that Solomonic wisdom applies to other primates: the mother gave up her infant only after a tussle and apparently did so rather than risk injuring it in a tug-of-war.

Thus, based on the limited evidence available for Abu, there is little to indicate that a sibling or other close relative would necessarily be more attractive than some other, younger, and equally available infant. Kinship is probably not a major factor in an allomother's choice of infants except insofar as relatedness may increase the allomother's access to a given infant. Females in those reproductive categories where it is advantageous for them to caretake will attempt to do so irrespective of how related they are to the infant and will sometimes take an infant without regard for the well-being of their charge.

Brutal Multiparas and Solicitous, Wary Nulliparas

The treatment dealt to infants by their allomothers ranges from care and solicitude to brutality. Whereas some allomothers treat the infant with the same solicitude that a real mother would (except that they do not suckle it), other caretakers are very nonchalant or even rough. The term "mistreatment" is used here to describe pulling, pushing, holding, dragging, or biting an infant in such a way that physical injury is a possibility and discomfort seems assured. Such mistreatment occurs mainly in five contexts having to do with getting the infant, keeping it, getting rid of it, inspecting it, and, on rare occasions, punishing it.

In many cases, the allomother chooses to take a noncompliant infant. Under natural conditions, the infant's prolonged separation from its mother or caretaker results in death; not surprisingly, an infant literally clings for its life to whichever female has it. Consequently, a prospective caretaker frequently has to pry or forcibly pull an infant off its current caretaker (color fig. 6).

Once an allomother has the infant, if the infant remains unwilling or fails to cling properly, it may be pulled along by one limb or by its tail or may be dragged by the allomother as it clings to her leg, arm, or tail. Of 14 instances in which infants were dragged, experienced females were involved 64 percent of the time, suggesting that the improper carriage of the infant resulted more from the allomother's unwillingness to inconvenience herself by stopping to pick up the infant than from her lack of practice. Of 244 allomaternal episodes in which experienced allomothers succeeded in taking infants, their charges were dropped on 6 occasions. On these occasions, infants were pulled along the ground by experienced females 4 times. Of 190 episodes in which inexperienced females took infants, their charges were dragged on 4 occasions. The only inexperienced female observed to pull infants along the ground was the three-legged nullipara, Guaca, who did so twice.

On average, nulliparous females dragged infants or pulled them along the ground on 3 percent of the occasions that they succeeded in taking them. These tallies indicate that infants suffered no more from being carried by females without previous maternal experience (although most of the successful nulliparas were older juveniles and subadults). If anything, nulliparas dragged infants slightly *less* often than multiparous females, who pulled or dragged their charges closer to 4 percent of the time. Occasionally, a nullipara would take off running before the infant had had a chance to properly catch hold, but more typically these skittish nulliparas would run bipedally for a short distance, with one or both free hands holding the infant to themselves while the infant caught hold. Of the 15 occasions when allomothers were observed to run bipedally while carrying an infant, 11 involved older nulliparas.

Other instances of improper carriage resulted when the infant either slipped or was pushed away from the allomother's chest and was forced to cling to some other part of her anatomy—the stomach, rump, sides, or legs (fig. 7.4). Of 113 such episodes, females who had already had one or more infants were involved 80 percent of the time.

Abuse of the infant may also result from a lactating allomother's attempts to divert a borrowed infant from her nipples, or else from an allomother's efforts to shed the infant altogether. To prevent an infant from nursing, an allomother may shove it down onto her stomach, push it onto the ground or pull it off and hold it upside down.

Allomothers may also push an infant off simply because they have tired of them (or so it appears) or because the infant's complaints and struggles have become a nuisance. The more that the allomother tries to rid herself of the infant, the more fiercely the infant complains, struggles, and clings tightly to whatever furry portion of the aunt's anatomy presents itself. To forestall this, the allomother sometimes pushes the infant off and then presses it face-down against the ground or a limb (fig. 7.5), or else she steps or sits on it (fig. 7.6). On the evening of the day that Brujo was born, I.E. pressed him against a granite boulder and stepped on him. The behavior of langur allomothers in these instances is not obviously different from the so-called "pathological" behavior dealt to infants by "motherless mothers" in experiments with rhesus macaque mothers who were raised in isolation (Harlow, 1971: fig. 3.3), and

Fig. 7.4. To keep from being abandoned, Brujo sticks like glue to T.T.'s rump.

yet, this is apparently normal treatment of infants among langur allomothers.

Of 34 occasions during the 1973-74 study period when infants were pushed off and held against the ground, almost all (94 percent) of this infant riddance was undertaken by older multiparous females. Only twice were nulliparas ever observed to try to push an infant off.

Despite the allomother's attempts to pry the infant loose, the infant is usually able to remain in tactile contact with the aunt. Hence, allomothers are sometimes seen walking along with a tiny baby clinging to their buttocks or the root of the tail (as in fig. 7.4). The precariousness of the infant's position, and especially its distress vocalizations, apparently advertise its availability to other allomothers and attracts them. It was my impression that allomothers

Fig. 7.5. I.E. presses day-old Brujo face-down against a granite boulder in an attempt to keep the newborn infant from clinging to her.

were much more likely to seek out a screaming infant than a quiet one. As a result, an infant would often be taken by another allomother before its current allomother succeeded in discarding it. On 10 occasions, however (and usually only after considerable difficulty), infants less than one month old were actually abandoned by their allomothers. That is, they were left on the ground, unattended by any langur. In all cases (summarized in table 7.8), deserted infants were retrieved within minutes; there was an equal likelihood that the infant would be picked up by its own mother or an allomother (fig. 7.7a-d). In 6 of the 9 cases in which the identity of the

Fig. 7.6. Guaca attempts to rid herself of a clinging infant by sitting on it.

deserting allomother was known, she was a multiparous female who was either pregnant and approaching term or else she had a recently delivered infant of her own.

Yet another type of mistreatment—one which is very rare—might be designated "punishment." It was my impression that an allomother in possession of an infant which was constantly struggling and which would not respond to attempts to calm it (by nuzzling or hugging the infant, or by scratching its back) might try to quiet the infant by punishing it. On 3 occasions, allomothers nipped a struggling, noncooperative infant; none were serious bites.

Finally, genital inspections might also be considered as a mild form of mistreatment; these involve holding the infant upside down, usually by the leg or tail. This treatment is more discomforting to the infant (which complains accordingly) than it is dangerous. Of 69 occasions when allomothers turned infants upside down, pregnant multiparas did so on 31 occasions, recent mothers on 23, experienced females without infants or with weaning infants on 8. Despite the high incidence of nulliparous allomothering, older nulliparas only turned infants upside down on 7 occasions. Younger nulliparas, under fifteen months of age, were never seen to turn an infant upside down to inspect its bottom (but they also had few opportunities to hold infants at all).

None of the observed forms of mistreatment, with the exception of trying to push an infant off, pressing it to the ground, or abandoning it, are unique to allomothers. Though it is rare, a real mother may occasionally carry her infant improperly or in such a way that the infant has to cling opportunistically to whatever portion of her body it can catch hold of. I once observed Kasturbia walk off with her infant Nuka sitting on top of her right haunch. As Kasturbia moved, Nuka toppled off and hung upside down and was momentarily dragged against the ground. Kasturbia then stopped and set her infant right. The only mother who mistreated her infant with anything like the incidence of an allomother was the handicapped female Guaca. Because it was difficult for this three-legged female to adjust her infant, Guat had to cling whichever way she could when her mother moved. On one occasion when Guat was in a particularly awkward position, Guaca pushed her daughter down with her foot much as a shedding allomother might do.

Although multiparous allomothers are rougher with infants than their real mothers are, this distinction only applies to the treatment of *new* infants, several months old or younger. After the sixth month or so, when weaning begins, and especially after about the thirteenth month when weaning is well under way, mothers may be ruthless about slapping an offspring away from the nipple, pushing it down or pressing it to the ground, or running away from it.

The fact that mothers treated their infants with greater solicitude than some allomothers is scarcely surprising. The fact that experienced females were as likely—if not more likely—to push, pull or drag infants, and that they were unquestionably more likely than inexperienced females to push an infant off, press it to the ground, or abandon it, was more startling, and led me to reexamine the

---

Fig. 7.7. (a) Mopsa holds Mole's newborn daughter, Moli (who clings to Mopsa's right leg), in addition to her own infant. (b) Mopsa pushes Moli off and departs, carrying her own son, but abandoning Moli. (c) Within a minute, the abandoned infant is approached by another allomother, T.T. (Note identifying purple stain on her head.) (d) "Rescue" is not always pleasant: T.T. performs a genital inspection by holding the complaining infant upside down.

TABLE 7.8. Summary of infant desertions by allomothers.

Date	Infant	Allomother who deserted	Apparent cause of separation	Female who retrieved	Time infant spent unattended
2/23/73	Niza	T.T.	Unknown	Mother	11:22 – 11:28
1/20/74	Moli	Mopsa	Negligent carriage	T.T.	15:29 – 15:31
1/22/74	Moli	I.E.	Not seen	I.E. (same female that abandoned)	12:09 – 12:09 +
1/22/74	Moli	I.E.	I.E. forces infant to ground	Mother	12:09+ – 12:11
1/22/74	Brujo	Handy	Infant crawls off on own accord	Mole	14:31 – 14:31 +
1/22/74	Moli	Mopsa	Mopsa pushes off Moli while continuing to suckle her own infant	T.T.	15:33 – 15:34
1/24/74	Moli	Niza	Probably negligent carriage	Pandy	8:14 – 8:14 +
1/28/74	Moli	T.T.	Infant crawls off on own accord	Mother	9:55 – 9:55+
1/29/74	Moli	Pandy	Unknown (aunt may have left infant in order to forage on ground)	Mother	11:03 – 11:04
1/31/74	Brujo	Unknown	Unknown	Mother	Retrieved ca. 8:53

Average time an abandoned infant was unattended: ≈ 1.8 minutes treatment of infants by individual allomothers and to score each female according to how often she mistreated infants (Blaffer Hrdy, 1975).

Allomaternal episodes were reviewed for the occurrence of any of the following rough behaviors: if the allomother yanked the infant off the mother; if the infant complained constantly throughout the episode or if the infant complained in a particularly frantic way; if the infant was pushed away from the chest and then held awkwardly or precariously to the allomother's stomach, rump, legs, or sides, or if the infant hung upside down once, or more times; and finally, if the allomother bit the infant. The total number of episodes in which an allomother was rough with an infant was then divided by the total number of times that she ever succeeded in taking or carrying the infant (see Blaffer Hrdy, 1975). When females in the five reproductive categories were ranked according to what percentage of the time their members had been observed to mistreat infants, younger nulliparas ranked first in roughness, having mistreated infants approximately 75 percent of the occasions on which they were able to obtain them. After these very young and inexperienced females came pregnant females (64 percent); recent, lactating mothers (52 percent); experienced females in the process of weaning an infant or else not currently associated with an infant (42 percent); and finally the older nulliparas in all troops (22 percent). In other words, except for nulliparas under fifteen months (who rarely succeed in taking infants anyway), experienced females who had had one or more infants, were far more brutal in terms of neglecting, mistreating, or eliciting prolonged complaints from infants. Of 405 allomaternal episodes in the Toad Rock and Bazaar troops recorded during ad libitum and focal-infant samples during the 1973-74 study period, inexperienced females were classified as rough on 25 percent of the occasions that they took infants (155 times). Experienced females were so classified over twice as often, 56 percent of the time (250 occasions), a difference which was highly significant. Older nulliparas were less likely to mistreat infants or to elicit prolonged complaints from them, and this was so even though on average older nulliparas kept infants much longer than other allomothers did (table 7.9).

In contrast to the rough treatment dealt to infants by multiparas, the interest of subadult females in their charges persisted over long periods of time and was characterized by solicitude for the infant's welfare. Subadult females appeared to be particularly concerned with keeping the infant from complaining. These nulliparas almost

TABLE 7.9. The number of minutes that allomothers in different reproductive categories held infants, based on 1973-74 data from the Toad Rock troop.

| | Nullipara | | | Pregnant | Experienced female | |
	Under 15 months	Over 15 months	Lactating mother	female near term	Weaning	No infant
Average number of minutes observed holding infant	2	9.97	3.93	5.08	3.14	7.16[a]
Number of episodes	7	108	68	102	7	6
Standard deviation	0.93	9.84	4.32	5.36	2.90	7.82
Range	1-3	1-36	1-20	1-31	1-10	1-24

a. This small sample was distorted by the inclusion of a single female who kept an infant 24 minutes.

never pushed an infant off themselves and only twice did older nulliparas abandon an infant.

Two behaviors in particular typify the solicitude and eagerness of subadults. These young females would grab an infant and take off running bipedally, clutching their charge against their bodies with one or both hands, as though they anticipated that some other female might take the infant from them if they hesitated even for a moment. The second nulliparous idiosyncracy was that of scratching the infant's head or back. This appeared to be a relatively effective means of soothing the infant. It was my impression that subadults kept infants quieter than did multiparas, and it seems possible that the persistent solicitude shown by subadults had the same goal as their wariness and their habit of taking the infant and running—to prevent retrieval of the infant. If a complaining infant is more prone to be retrieved by another allomother or by the real mother, then both mother and allomother would serve as referees who respond to mistreatment of the infant by taking it away from the offending allomother. As long as a female is motivated to retain the infant, it behooves her to keep it quiet. Lancaster (1971) has argued that such conditioning helps the nullipara to learn appropriate maternal behaviors. Among the vervet monkeys she studied, the infant's cry immediately brought the mother to rescue it; if her off-spring was being abused, the mother would bite the caretaker. The brutal treatment of the kidnapped School troop infant described above and the fact that a 10-month-old infant, Nonet, who would not normally be allowed access to a days-old baby, was nevertheless able to take this kidnapped infant underscore the importance of maternal presence for supervision among langurs as well.

Unlike a vervet, a langur mother may ignore the most desperate cries from her infant if she is otherwise occupied. Only if her infant is endangered—by the approach of a dog or human, for example—would a mother interrupt her feeding to retrieve it. As a nullipara in my own right, this nonchalance on the part of mothers and the heart-rending screams of their infants made it stressful for me to concentrate on infants for hours on end. It was with relief that in January of 1974 I laid aside the problem of allomothers to move on to other topics. I felt that I had found answers to many of the questions originally posed.

Several findings about allomothers are straightforward. Females who have not yet given birth make the most eager caretakers. Whereas younger nulliparas—inexperienced and immature females under fifteen months of age—make the roughest allomothers, nullip-aras older than fifteen months are the most solicitous and their interest the most sustained. These facts are completely consistent with basic premises of the learning-to-mother hypothesis. With practice, females improve their maternal skills; hence nulliparas who have the most to learn will have the most to gain from allo-mothering.

Other findings, however, are not at first glance consistent with the learning-to-mother hypothesis. These seemingly counterintuitive findings deserve some further explanation. Curiously, older and more experienced females who have already had several infants make rougher allomothers than do some nulliparous females. The roughness of the treatment dealt to infants by experienced females rules out, in most instances, the possibility that such caretaking be-havior could be either altruistic or investment in matrilineal rela-tives. The infant's mother, however, may benefit from temporary freedom to forage. It is interesting to note in this respect that, gen-erally speaking, other adult females will be more closely related to the mother than to her offspring (see p. 47). Kin selection is prob-ably not at issue, however; infants do not appear to be either chosen or avoided on the basis of degree of relatedness. Nor would kin selection explain why an experienced and supposedly competent female would take an infant and then mistreat it. Furthermore, this mishandling cannot be written off as spite nor as an attempt by one adult female to rid her own offspring of future competition, since in fact infants are only rarely killed or disabled by allomothers.[6] The behavior of allomothers is best described as lacking solicitude, not as malicious.

If the mistreatment measured in this study is in fact a reflection of allomaternal competence, that finding would have to be in conflict with the learning-to-mother hypothesis. But actually I do not believe that maternal skills are at issue. Rather the key to the curious contrast in the treatment dealt to infants by experienced and inexperienced females may lie in an additional finding: whereas younger females rarely try to discard an infant, experienced females may begin to abuse their charge and even to attempt to discard it soon after taking it. Many of the behaviors which were classified here as brutal or rough (pushing an infant off; pressing it to the ground; failing to pick up an infant that is being dragged) are outgrowths of allomaternal attempts to get rid of infants. It is this conflict of interest between an allomother who wishes to shed an encumbrance and an infant that clings which underlies the brutality that was being recorded.

Both the difference in the motivation of experienced and inexperienced caretakers to take infants, and the different lengths of time that each category is motivated to keep infants, might be explained by the following two-part hypothesis: If a nulliparous female without the prior experience of rearing an infant gains increased competence the longer that she holds and carries an infant, and if she does not soon reach the limit where further practice would be redundant, such a nulliparous female would be motivated to keep a newborn infant for as long as she can and still fulfill her requirements of foraging and rest. Where there is competition between nulliparas for access to infants, she would rarely have access to an infant long enough for her practice to become redundant. On the other hand, females who have had a previous infant and who are being primed for motherhood or having their memories refreshed, or who are simply gathering information about a new troop member, would not gain to the same extent from sustained contact with an infant. Rather, in the case of multiparas, the experience of caretaking

6. There is little question that a truly malicious female has the capacity to dispatch an infant in her charge. According to S. Ripley (personal communication), adult females may sometimes attack and kill an infant from a neighboring troop.

soon becomes redundant and at that point allomothers may either ignore the infant (which must then cling as best it can) or else actively try to remove it. In addition, experienced females would have less incentive to keep an infant from complaining or to otherwise forestall its retrieval.

If these suggestions are correct, mistreatment of infants by experienced females is only indirectly related to the learning-to-mother hypothesis and for this reason cannot be said to detract from its validity. The extraordinary solicitude and competence exhibited by many older nulliparas, however, offer direct support of it. The short duration of interest in infants and the consequent brutality to infants exhibited by multiparous females may have important implications for the two supplementary explanations for allomaternal caretaking: priming and information gathering. The relationship between these two types of motivation to caretake and the observed short interest span can best be explained by taking into account the concept of redundancy, that is, there would be a limit to how much priming and informational input would be valuable, and this ceiling would be reflected in the short span of a female's motivation to hold an infant.

Cooperation and Exploitation

Fifty percent of the genes in any given population of primates derive from females, and male fitness is entirely dependent on extensive and long-term investment by females. Nevertheless, with the exception of maternal behavior, female reproductive strategies have been largely ignored in field studies of primate social behavior. The following statement from an account of langur social behavior carries to an extreme what has been a general attitude within primate studies: "The dominant position within the society is held by the adult males; the number of adult males and their reciprocal relationships determine the social structure of the group as well as the group behavior as a whole" (Vogel, 1973a:363). However, the *Weltanschauung* which inspired such interpretations is changing.

The earlier bias, now imbedded in the literature and in the minds of the students who read it, tended to obscure the core role played by females in primate social organization. Matrilineal relationships are the basic bond for long-term social relationships among the best-studied species of baboons, macaques, orangutans, langurs, vervets, and some prosimians, as shown by the recent work of a number of primatologists including J. Lancaster, G. Hausfater, P. Rodman, D. Sade, and R. D. Martin. Though the behavior of males may have a profound effect on the reproductive success and strategies of females, adult males figure marginally, if at all, in such basic female pursuits as competing with one another and learning to mother. Furthermore, the female adaptation is often the prior one. Females are dispersed about the environment in accordance with their foraging and child-rearing needs; male strategies co-evolve so as to exploit this distribution.

Among langurs, the presence of males is crucial not only for insemination but also for the defense of the troop against other males. Nevertheless, many female strategies like those of males are largely intrasexual. In order to highlight the importance of female-female interactions, competition and cooperation between females have been analzyed in this study with little regard for males, even though this omission creates a somewhat artificial impression—especially in the area of dominance relations. Normally an adult male can displace any female in the troop regardless of her rank within the female displacement hierarchy. Whatever a female gains in terms of access to resources from high rank within the troop, males almost always have prior access. Because of their large size and fighting ability, males have a physical advantage at any point where a conflict of interest between the sexes occurs.

Although males may benefit from the acquisition of maternal skills by future consorts, allomothering is out of their sphere; males play no part in what is essentially an exploitation of newborn infants to serve the ends of troop females. Langur allomothering disrupts nursing, subjects infants to physical discomfort, and threatens them with the possibility of damage or desertion. Because the fate of a newborn infant is almost totally in the hands of older animals, it is not essential that the infant benefit to the same degree that mother and allomothers do in order for the behavior to take place. The mother, however, must acquiesce, since the option remains open to her to either keep hold of her infant, to stay clear of potential caretakers, or to leave the troop—as seriously threatened mothers sometimes do when confronted with an infanticidal male. The fact that allomothers will on average be more closely related to the infant's mother than to the infant may provide additional grounds for cooperation between the mother and allomothers, but insofar as the infant is concerned, caretaking is more nearly exploitation.

Not surprisingly, there is little evidence that allomothers seek out close kin; they may even prefer more distantly related infants stolen from neighboring troops. In this respect, langur allomothers differ from so-called "helpers at the nest" among the numerous bird species in which offspring of previous years defer breeding to assist in feeding and tending siblings at the parental nest (Skutch, 1961; Woolfenden, 1975). To the extent that nulliparous and pregnant primates gain from practice and priming for motherhood, these allo-mothers are pursuing a primarily selfish rather than kin-selected strategy. An even more extreme exploitation of infants is described in the next chapter.

8 / Infanticidal Males and
   Female Counter-strategists

We need [men] sometimes if only to protect us from
other men.

Alison Lurie, 1974
The War between the Tates

In the early afternoon of August 12, 1972, the Hillside troop feeds quietly in the trees lining the drive of the Phiroze school. There is an undercurrent of tension, however. Very deliberately, females in the troop avoid Mug. If Mug climbs into a tree where a female is feeding, she immediately moves to another tree. If Mug follows, the female returns to the tree where she was before. At 4:00 p.m., the male, who is grunting softly, climbs to the top of a nearby roof. From his high vantage point, Mug strikes a sentinel's pose and stares off into the distance. Then, abruptly, the stocky gray form descends from his rooftop perch, charges directly at Itch, and grabs at the infant clinging to her belly. Itch whirls to face Mug, plants her front paws as she lunges at him, grimaces, and bares her teeth. Within seconds of this assault, old Sol, the one-armed female Pawless, and a third, unidentified female join in the counterattack. The three defenders interpose themselves between Mug and Itch, lunge at the male, and chase him up a tree. In the wake of this assault, a trembling Itch is left alone to hold her infant Scratch, who is spattered with flecks of blood.

Here was the phenomenon, the bizarre Dharwar aberration of adult males attacking infants, that had brought me to India to study langurs. But I watched these events with the same incredulity and anxiousness that I had felt the year before, during the unusually rainy monsoon of August 1971, when I had first realized that Shifty Leftless, the male who had just usurped control of Hillside troop, had probably killed the six infants present in the troop at that time. Although infanticide was foremost on my mind when I decided to study langurs, its actual occurrence seemed totally implausible. Despite Shifty's replacement of Mug, despite the fact that all six infants were missing, despite reports by two local people who had seen an adult male langur kill infants in the Hillside troop's home range, I grasped at straws. I spent a whole day trying to convince myself that this was a different troop, one without infants, which had somehow materialized out of the torrential rains and thick mist of that monsoon month. But the longer I peered through the mist at those rain-soaked, skittish females, the more I realized they were, unmistakably, Bilgay, Itch, Harrieta, Oedipa, Pawless, and Sol. The seventh female and the six infants were never seen again.

It is apparent by now that these events were not merely an aberration. Increased observation of primates by scientists has meant an increase in the known number of species in which adult males attack and kill infants. Some of these incidents are clouded in mist, others leave little room for the imagination. In his field report of the langurs of Jodhpur S. M. Mohnot writes:

> On 24 July 1969, about 8:20 a.m., the new leader (YA male-1) was observed sitting on the spire of the North Temple, 'muttering' slowly. The females were sitting on the hillock about 10 metres away from him in the shade of the temple. At 8:33 a.m., L female went up to the leader and initiated grooming, but the latter did not offer any part of his body for grooming. When she tried again, she was threatened . . . The male then started grunting; at the same time he started staring intensely in all directions and strolling on the spire. This was followed by the sharp grinding of his canines. Facially he looked tense and gave a whoop call twice and this was accompanied by the full erection of his penis . . . Then the male produced a rare deep-throated alarm bark in high pitch (hiicheeik-heikhe, hiicheeik-heikhe . . .) . . . He continued the grinding of canines interrupted by 'groaning'.
>
> About 9:50 a.m. the male, with a sudden bound, was among the females. He grabbed the infant from the lap of Ti female, clasped it in the right arm, held its left flank in his mouth and ran fast towards the northern periphery of B-26 home range. The mother (Ti female) and two other females . . . rushed at the running male. The mother twice blocked his passage but could not recapture the infant; the other two females also failed. All the while the infant was screeching (cheeen, cheeen, cheeen . . .) After running for 70-80 metres, the male stopped for a moment, took a quick bite at the infant's left flank with his canines [producing a 6-centimeter gash through which part of the infant's intestines emerged], dropped him on the ground and sat near

the bleeding infant. When the mother approached him, he barked loudly, tossed his head, bared his teeth and stared at her. She was apparently frightened and stayed away at a distance . . . The whole process, from the time the male snatched away the infant and his subsequent dropping it, took 3 minutes. After staying for 5 or 6 minutes near the infant, the male left him in a dying condition and went under a *Euphorbia* bush about 6 metres away. (Mohnot, 1971b:181)

Infanticide actually witnessed, or the suspicious disappearance of a cohort of infants at the time that a new male entered the troop, has been reported for some dozen species of primates. Not all these reports parallel the pattern of events recorded among hanuman langurs. But many are disturbingly similar: males attack infants after gaining access to, or possession of, a female whose offspring was sired by some other male. Usually, this is a female previously unfamiliar to him. Perhaps the clearest example of the potential importance of previous acquaintance with the female is provided by a study of caged crab-eating macaques. In this instance, infanticide was the unexpected outcome of an experimental study of the effects of familiarity or lack of it among opposite-sex pairs of *Macaca fascicularis*. When paired with his usual companion and her infant, the adult male displayed typical behaviors: mounting the female briefly and then casually exploring his surroundings. He entirely ignored the infant. Paired with an unfamiliar mother-infant pair, the male responded quite differently. After a brief attempt at mounting, the male attacked the infant as it lay clutched to its mother's belly. When the mother tried to escape, the male pinned her to the ground and, gnawing the infant, made three different punctures in its brain case with his canines (Thompson, 1967; see also Washburn and Hamburg, 1968).

Two suspected cases of infanticide among wild hamadryas baboons in Ethiopia also appeared to have been occasioned by human manipulation. In the course of capture-and-release experimentation on the process of harem formation among *Papio hamadryas* in Ethiopia, two mothers with infants changed owners. In one case the infant was missing a day later; in the other, the infant was seen dead, its "skull pierced and its thighs lacerated by large canine teeth" (Kummer, Gotz, and Angst, 1974). The killing of two hamadryas infants witnessed at the Zurich Zoo just after their mothers had changed hands adds plausibility to the suggestion that the wild infants were similarly murdered.

Less contrived perhaps is an account of chimpanzees from the Gombe Stream Reserve in Tanzania. British researcher David Bygott was following a band of five male chimpanzees when they encountered a strange female whom, "in hundreds of hours of field observation," Bygott had never seen before. This female and her infant were immediately and intensely attacked by the males. For a few minutes, the screaming mass of chimps disappeared from Bygott's view; when he relocated them, the strange female had disappeared, and one of the males (Humphrey) held a struggling infant. "Its nose was bleeding, as though from a blow, and Humphrey, holding the infant's legs, intermittently beat its head against a branch. After three minutes, he began to eat the flesh from the thighs of the infant, which then stopped struggling and calling" (Bygott, 1972; see also Suzuki, 1971). In contrast to normal patterns of chimp predation, this cannibalized corpse was nibbled by several other males but never consumed.

In each of these incidents, adult males killed infants belonging to strange females. A number of field workers during the last five years have reported the disappearance or mauling of infants at times when adult males entered the group from outside it. Perhaps the most dramatic of such instances was one of three recorded by Dian Fossey among wild mountain gorillas (*Gorilla gorilla beringei*) of central Africa. For several days, a lone silverback, or fully mature, male had been following a harem of gorillas, presumably in quest of females. At last he made his move and entered the group with a "violent charging run." A primiparous female that had given birth to an infant on the previous night countered his charge by running at him. Halting within arm's reach of the male, she stood bipedally to beat her chest. The male struck her exposed ventral body where her newly born infant was clinging. Immediately following this blow, a "thin wail" was heard from the dying infant (Fossey, 1976; 1974). On two other occasions, Fossey observed the infants of primiparous mothers killed by silverbacks from other social units. In the best-documented of these cases, the mother subsequently copulated with the male who killed her infant (D. Fossey, personal communication).

The pattern of male take-overs accompanied by the disappearance or killing of infants is especially prevalent among members of the subfamily Colobinae; they have been reported for wild silvered leaf monkeys in the forests of Malaysia, purple-faced leaf monkeys in Sri Lanka, the Mentawei Island leaf monkeys, and among the black-and-white colobus monkeys of Uganda (see p. 17). Outside of the Colobinae, the take-over/infanticide pattern has now been reported among *Cercopithecus ascanius*, the arboreal redtailed monkey of central Africa (Struhsaker, 1976).

In addition, isolated instances of infanticide committed by adult males is known to occur among various species of prosimian (Mitchell and Brandt, 1972), the Barbary macaque (*Macaca sylvana*, Burton, 1972), and free-ranging rhesus macaques (Carpenter, 1942; D. Sade, personal communication); but the circumstances surrounding these attacks are incompletely known. Infanticide is suspected among wild chacma and anubis baboons (Saayman, 1971; J. Moore, personal communication), wild howler monkeys of South America (Collias and Southwick, 1952), and among caged squirrel monkeys (Bowden et al., 1967). Only among hanuman langurs, however, is there sufficient information to say that infanticide occurs regularly, and under conditions that must now be considered normal for this species because these conditions are both widespread and of long duration. It makes sense, then, to turn to the Hanuman for elucidation of this chilling phenomenon.

Several explanations have been offered. Most of these derive from the underlying assumption that under normal conditions animals act to maintain, not disrupt, the prevailing social structure; behavior that upsets this balance must be temporarily "pathological," and social pathology has figured prominently in the explanations for infanticide. A number of writers have suggested that infanticide is a product of high population densities (Sugiyama, 1967; cited in Crook, 1970; Eisenberg et al., 1972), and as such is a possible mechanism for population control (Rudran, 1973b; Kummer et al., 1974). Alternatively, it has been suggested that the behavior has no adaptive value (Bygott, 1972; Dolhinow, 1977; Curtin, 1977) and that the behavior is pathological or "dysgenic" (Warren, 1967). Functional explanations for infanticide include the idea that males are somehow displacing aggression built up by the "simultaneous sexual exitement and enragement" of the new leader (Mohnot, 1971b), or that the male attacks in order to impress and to strengthen his "social bonds" with the females in his new troop (Sugiyama, 1965b; cited in Spencer-Booth, 1970).

Only one of the early explanations focused on the possible advantages of infanticide for the animal actually responsible for the act—the male. In 1967, Sugiyama suggested that the male was attacking the infant so as to avoid a two-to-three-year delay in female sexual receptivity should the mother continue to nurse her infant. This argument can be expanded into a more general interpretation invoking sexual selection (Blaffer Hrdy, 1974). As defined by Darwin (1859) sexual selection refers to any struggle between members of one sex (typically males) for access to the other, with the result for the unsuccessful competitor being not death but few or no off-spring. According to this hypothesis, infant killing is part of a reproductive strategy whereby the usurping male increases his own reproductive success at the expense of the former leader (presumably the father of the infant killed), the mother, and her infant. As such, infanticide would be an extension of competition among males, a phenomenon occurring in a wide variety of birds and mammals—wherever females invest substantially more than do males in the production and rearing of offspring (Trivers, 1972).

In the remainder of this chapter I will detail the evidence concerning infanticide among hanuman langurs; I hope to demonstrate that the evidence supports a sexual-selection hypothesis far more convincingly than it does any of the competing explanations.

Struggle for Control of the Hillside and Bazaar Troops

In July 1971, Hillside troop consisted of a single adult male (Mug), six adult females, including the three-legged female named Pawless, six infants (including the pair of twins), a juvenile male (Sancho) who may have been Pawless' son, and a very old female (Sol). When reencountered on August 5, Shifty Leftless, the permanently identifiable male with a portion of his left ear missing, had taken over the Hillside troop. All six infants and one adult female were missing. Between August 4 and August 16, three of the mothers (Bilgay, Itch, and Oedipa) came into estrus and copulated with Shifty (fig. 8.1). During this period neither Pawless nor Sol, who did not have infants at the time of the take-over, came into estrus. Both females, but especially Pawless, harassed the new alpha. Pawless would charge repeatedly at Shifty; occasionally other females would join her, and three or more females might be seen chasing Shifty.

Shortly after I first noticed the presence of the new male, I learned from local inhabitants that an adult male langur had killed two infants, one near the Phiroze School and one near Rajendra Road, which runs past Boulders.[1] These sites were both located in areas used intensively, and in the case of the Phiroze, exclusively, by Hillside troop (fig. 4.5). This information was in accord with observed changes in troop composition and with the behavior of the females, as well as consistent with similar accounts from Dharwar and Jodhpur (Sugiyama, 1967; Mohnot, 1971b). The concurrence of

---

1. I am grateful to Mrs. Phiroze Merwanji and to an unidentified police cadet for these observations.

Fig. 8.1. Shortly after his take-over of Hillside troop during the monsoon season of 1971, Shifty copulated with females whose infants had disappeared. One estrous female waits in foreground while Shifty copulates in the mist with her troopmate.

these diverse clues led me to assume that the new male had killed at least two, and probably all six, infants.

In the first weeks after the new male entered the troop, Pawless stayed at the periphery of the troop with Sancho, only approaching the center in order to initiate harassment of Shifty. By August 21, however, Pawless herself came into estrus and left Sancho on the outskirts of the troop while she joined Shifty in a consort relationship. Their copulations were subject to violent harassment from two of the former mothers, especially Harrieta. On one occasion when Harrieta interrupted a copulation, Shifty dismounted to chase her, only to be repulsed and then chased by every female in the troop except Pawless. On August 25 when Harrieta was in estrus, her copulations were similarly harassed by other females.

As far as I know, Shifty was the only male that copulated with these females after the take-over. However, Hillside troop was not under constant observation, and other males were sometimes in the vicinity of the troop. On August 20, an estrous Harrieta twice approached the leader of the neighboring I. P. S. troop, only to be chased and headed off each time by Shifty. On August 27, Mug attempted to reenter his former troop and in fact chased Shifty some distance before Shifty turned on him and chased Mug.

Around March or April of 1972, during my absence from Abu, a new male in Bazaar troop was detected and the killing of two Bazaar troop infants by a male was observed by an amateur ornithologist whose rooftop (Sanand House) was used by Bazaar troop as an overnight sleeping site.[2] When I encountered Bazaar troop on June 25, 1972, I was able to identify the new male as Shifty Leftless (fig. 8.2). Three infants (born late in July and in August of 1971) were missing, as were two older juvenile males, three adult males, and one adult female. On July 1, I encountered the second ranking of the former Bazaar troop males sitting in a tree near Sanand House nursing a deep unhealed wound in his left deltoid.

Between July 5 and July 16, six Bazaar troop females came into estrus and copulated with Shifty. Seven months later, in February of 1973, two of these females were accompanied by 3-to-4-month-old infants, two carried newborns, and one was in the last weeks of pregnancy. The sixth female gave birth eight months later, around October of 1973. Using six to seven months as the probable gestation period and counting backward, it seems that two of the females conceived in the period from April to May and were already pregnant during the copulations that I witnessed. In the case of two other females, what I saw may have been the second or third cycling since the death of their infants, and they may have conceived on this occasion. The fifth female probably conceived a month or two later.

Based on observed demographic changes, the witnessed killing of two infants, and the location of those killings, I concluded that around April, Shifty had left Hillside troop for Bazaar troop, had

2. I am grateful to Nirmal Kumar Dhadlal and to his daughter, Gaitry, for these and other observations.

Fig. 8.2. Shifty Leftless sits surrounded by his new Bazaar harem in June 1972. (Note Shifty's ragged left ear.)

evicted the three resident males, and had been responsible for the deaths of the missing infants.

Even after Shifty took over Bazaar troop, he attempted to retain exclusive access to Hillside troop; throughout the monsoon season of 1972, Shifty and Mug vied for troop control. Whenever Hillside troop approached their common boundary with Bazaar troop, if Shifty and his new troop were in the vicinity, Shifty would leave Bazaar troop and temporarily join the Hillside females. Occasionally, Shifty would leave Bazaar troop and travel some distance (as much as one kilometer) to check on his former harem. On two such occasions, Shifty whooped several times prior to his actual arrival. This apparently investigative call was never answered by Mug; rather, it served as a warning to Mug, who promptly left the vicinity. On five occasions, Mug faded away before Shifty actually showed himself. On three other occasions, however, Shifty arrived unannounced and Mug decamped with Shifty in pursuit.

On these occasions, Shifty spent an average of about two hours with Hillside troop. The short duration of his stays may have been due to the dearth of estrous females in that troop. On August 5, 1972, when Pawless was in estrus, Mug was chased out twice (once at 9:30 a.m. and again at 5:30 p.m.) and on that day Shifty spent a total of eight hours with Hillside troop, mostly in consort with Pawless.

On days when Bazaar troop remained in close proximity to Hillside troop, Shifty stayed longer, but at such times it was difficult to say which troop he was actually with. As far as I know, Shifty always returned to Bazaar troop before nightfall. On five occasions in 1972 and 1973, however, Hillside troop spent the night in trees or on a rooftop adjacent to Bazaar troop's sleeping site following days the two troops had spent together.

As many as three days might pass before Mug returned to Hillside troop. On one occasion, however, when Hillside troop had moved out of the vicinity of the Bazaar troop, Mug returned less than an hour after Shifty had left. Another time, Mug apparently misjudged the advisability of return. He loped down a road toward the Hillside females only to whirl and run away again at full speed when he was still some thirty meters from the troop: Shifty and Bazaar troop were on a hilltop only a short distance away.

The struggle for control of Hillside troop was still ongoing when I returned in the dry season months of February and March, 1973. At that time, Mug was accompanied by five adult males that I had never seen before. Whether these temporary joiners were with the Hillside females or apart from them, Mug remained individually dominant to each of the other males for access to position and to food (table 8.1). It is not known, however, if Mug's dominant position meant that he also controlled access to ovulating females. During 98 hours of observation when the males were with the females, Mug was never seen mounting a female. On only one occasion did he even display special interest in a soliciting female. On March 20, when Sol was in estrus, Mug was seen in consort with her in a dense thicket out of view from other males and out of my sight as well. Given the circumstances, undetected copulations could have occurred.

Even though Harrieta, Oedipa, and Itch all displayed estrous behavior (they presented to males and shook their heads) during this period, as far as I could determine, Sol was the only Hillside troop

TABLE 8.1. Displacements among Hillside males over positions and preferred food items.

Rank	Male	Mug	Righty	No-No	Kali	Blue-beard	Pequeno	Total
1	Mug	—	12	3	5	4	4	28
2	Righty	0	—	3	13	8	5	29
3	No-No Man[a]	0	1	—	0	0	1	2
4.5	Kali	0	0	0	—	0	0	0
4.5	Bluebeard	0	0	0	0	0	1	1
6	Pequeno	0	0	0	0	0	0	0
							Total:	60

a. No-No Man avoids direct confrontation with Mug, Righty, and Kali but is frequently tolerated by these males in situations where any other male would be
threatened away.

female who was actually cycling. This assessment is based on
detection of menstrual blood, intensity of estrous solicitations, and
increased aggressiveness, which typically accompanies estrous
behavior. Sol's menstruation was visible on February 12. On February 19, and again from March 17 to 20, she vigorously solicited
males. On these occasions she was mounted by Bluebeard (once),
Kali (twice), and Righty (four times); and, as mentioned above, she
was in consort with Mug on the afternoon of March 20th. Since Hillside troop was only observed briefly (between 30 minutes and two
hours) on March 18, 19, 21, 22, this list is undoubtedly incomplete.
Although Sol appeared to be cycling, it is not certain whether or not
she was actually fertile, since she was never seen with an infant. In
any event, she was the only female whose solicitations Mug
responded to.

All of the other males, except No-No Man, responded to the frequent but unusually mild estrous solicitations of Itch and Oedipa.
During February and March, these two females were mounted on
more than 45 occasions by Righty (31 times), Kali (6), Bluebeard (6),
and Pequeno (2). Mug never overtly interfered with these
copulations, but harassment by other males was intense. Righty
mounted Sol, Oedipa, and Itch on 47 occasions and was harassed
during 38 of these. The low-ranking male Bluebeard was responsible for 71 percent of this attempted interference. Though No-No
Man never participated in copulations, he joined in the harassment
of other males on 11 occasions.

Given that Mug was dominant to these newcomers and that I

never saw any evidence suggesting that they formed alliances among themselves, why did Mug allow the five males to remain? One possibility is that Mug tolerated the presence of these males in order to dissuade Shifty from his periodic visits. This view was supported indirectly by two forays into Bazaar troop's range made by the six males. On February 22, all six males, with Mug in the lead, approached within 60 meters of where Shifty and his Bazaar troop sat, on the other side of a low gully. This confrontation was preceded by whoops on both sides, and throughout it Mug and his band ground their teeth and grunted. On February 28, the six males made a noisy early morning expedition into the neighboring troop's range at a time when I knew (and I presume that the males did also) that Shifty and his harem were at Sanand House, at the furthest corner of their range from the males' invasion.

If Mug was in fact tolerating alien males in order to keep Shifty out of Hillside troop, the strategy was unsuccessful: Mug in 1973 may have been more brazen in the face of his antagonist, but whenever Shifty actually approached, the six males fled. During 334 hours that Hillside individuals were kept under observation in 1972 and 1973, the proportion of daytime that Mug was able to spend in Hillside troop did not change despite the presence of five extra males in 1973.[3] In 1972, Mug was with Hillside troop for 118 of 172 hours (66 percent), while in 1973, he spent 98 of 156 hours (63 percent) with the females. During this same period, Shifty was observed with the troop about 20 percent of the time, while the females (not including Pawless' 1973 wanderings which are discussed below) traveled on their own without any male present about 15 percent of the time. These figures do not represent absolute proportions of time that the males were either present or absent from the troop because of the following bias: I would observe them for a longer period if Mug was with the troop than if he was not. In my estimation, the amount of time that Mug actually spent with the troop was around 10 percent lower, closer to 55 percent of the time, while the figures for females traveling without a male should be greater, around 25 percent of the time. In any event, Mug did not seem to improve his position by the addition of the new males, since the bias presumably acted equally in 1972 and 1973.

Shifty's superiority was strikingly apparent when, on February

3. However, in 1973 Mug was no longer spending the night with the Hillside troop. Instead, he accompanied the five new males to one of two sleeping sites northeast of Hillside troop's home range.

15, 1973, all six males were dozing in a fig tree across from Hillside
House; Bilgay, Itch, and Oedipa were feeding nearby. Suddenly,
without a sound, all six males leapt from the tree and ran to the
north. I looked over my shoulder to find Shifty standing on a wall,
twenty yards from the base of the fig. More remarkable than the six
males' reaction to his presence was Shifty's utter silence, totally un-
characteristic of a langur male approaching foreign males. After a
moment's pause, Shifty took off to the north in pursuit of the six
males; I did not see any of them again that day. On four other
occasions, Shifty displaced the six males from Hillside troop, at
10:00 a.m. and again at 3:55 p.m. on March 3, at 1:50 p.m. on
March 6, and around 3:50 p.m. on March 13 (fig. 8.3).

Invariably, Shifty returned to Bazaar troop, and he spent the bulk
of his time with this larger harem. Shifty's absenteeism had import-

Fig. 8.3. Mug and three of the invaders sit on a hilltop across a valley from
Hillside troop's home range after having been chased out by Shifty Leftless.
Mug, in foreground, grinds his teeth.

ant consequences for mothers and infants in the Hillside troop. Throughout August and September of 1972, Mug stalked Hillside mothers with infants whenever he was with the troop. On nine occasions, a mother-infant pair was actually contacted by him, and in three of these assaults, the infant was injured. In six of these witnessed attacks, Itch and her infant Scratch were the target. In the other three, no definite identification was made, though in one case the target was probably Bilgay and her dark infant, Mira.

Nine assaults observed in 1972 are summarized in table 8.2. In every case, females other than the mother intervened. Sol and Pawless repeatedly played the most daring roles in defense of the infant (fig. 8.4). These same two females also interposed themselves between the male and a mother-infant pair on numerous other occasions, when Mug attempted to approach them. The following description of Mug's stalking, his thwarted assaults, and the intervention by other females is based on my field notes for August 12, 1972; these events took place after the attack on Scratch described at the beginning of this chapter.

Mug has retreated to a nearby tree; he stares down at the pursuing females and begins to bark ("agh, agh-agh") and to utter convulsive threats (a rare and peculiarly violent male vocalization composed of a sequence of spasmodic brays, probably the same sound Mohnot described as "hiicheeik-heikhe"). Moments later, Mug descends and again charges Itch. Almost simultaneously with his charge, Sol intercedes. She lunges at the male, slapping and contacting him with her forepaws. Mug returns to his tree and resumes barking.

During these events, Bilgay carrying Mira and Harrieta carrying Harry move to a tall *Grevillea* tree nearby. They are joined there by Oedipa with her infant and Sol. Momentarily the females relax and begin to feed. At 4:18 p.m. Mug retreats from Sol into the tree to which Itch and the other females have moved. Mug barks and then moves toward the females, grunting. They scatter again except for Sol, who climbs up the tree where Mug is and, for the third time, charges him. Pawless comes up behind Sol, then both females leave the tree. The male barks continually at three-second intervals for about ten minutes. At 4:30 Sol and the mothers disperse into three trees adjacent to the tree where Mug is.

At 4:40 Mug moves to the furthest of the three trees, where Itch is. She moves approximately five meters away, toward the end of the limb she is on, while Bilgay moves to the bottom of the tree. As Itch moves to a branch below Mug, he first crouches and then leaps down on her. Itch evades him, and Mug gives up temporarily.

At 4:47 Mug throws back his head and whoops; both Itch and Bil-

TABLE 8.2. Summary of witnessed assaults by Mug upon Hillside infants in August and September of 1972 during the period in which Mug and Shifty were vying with one another for control of Hillside troop.

Date	Time	Identity of infant	Identity of defending females	Male vocalizations	Location	Outcome
8/10	2:40 p.m.	Unknown	Unknown	Teeth grinding, grunts	Church	Attack thwarted
8/12	4:18-5:40 p.m.	Scratch	Pawless, Sol, Itch, and unknown female	Continuous barks and convulsive threats	Phiroze	Scratch superficially wounded
8/13	9:10-11:00 a.m.	Scratch	Sol and Pawless		Phiroze	Attack twarted
9/1	9:55 a.m.	Scratch	Sol engages Mug	Grunts	Eagle's Nest	Attack thwarted
9/1	3:30 p.m.	Unknown	Sol and Pawless engage Mug	Grunts, barks, and convulsive threats	Hillside	Attack thwarted

9/1	5:50 p.m.	Scratch	Sol and Pawless engage and chase Mug	Continuous grunts	Boulders	Scratch superficially wounded
9/5	4:45 p.m.	Unknown	Sol and Pawless	Soundless	Hillside	Mug grabs infant before thwarted
9/5	5:18 p.m.	Scratch (falls from tree)	Sol and Pawless engage, grapple with, and then chase Mug[a]	Soundless	Hillside	Mug grabs infant before thwarted
9/9	6:30 p.m.	Scratch	Sol and Pawless contact, slap at, and chase Mug[b]	Continuous barks; convulsive threats; whoops; teeth grinding[c]	Phiroze	Scratch severely wounded in thigh

a. See fig. 8.5.
b. See fig. 8.4.
c. Moments after the assault, the neighboring I.P.S. troop's leader appeared and chased Mug. He may have been incited to do so by Mug's threatening vocalizations.

gay leave the tree Mug is in. As the females pause for a moment on the ground, Itch's infant tries to leave his mother, but she restrains him. By 4:55 the two mothers have settled themselves in another tree. Mug approaches Bilgay, displacing her, and then moves slowly toward Itch and Scratch. He climbs to a branch just above them and begins to feed.

As Mug calms down, Itch allows Scratch to move about a meter from her; he returns within one minute. Moments later, Mug moves toward Itch, who continues to eat the branchlet she is holding. Mug stands nearby, staring at Bilgay and at Harrieta, though occasionally he looks over his shoulder, very briefly, at Itch, who seems to be the true focus of his attention. At 5:00 Mug leaps onto a branch directly above Itch's head; she does not interrupt her vigorous feeding until seconds later, when she moves to another tree.

Once again, Itch lets her infant wander a meter from her. As she does so, Mug's attention is riveted to Scratch. Mug moves slowly down from the tree he is in. Itch gathers up her infant and climbs higher in their tree. At 5:05 Mug climbs into Itch's tree. Itch and a second, unidentified female who is there leave abruptly. As Mug continues to stalk the mothers, following Itch and Bilgay out onto a limb, Sol accosts Mug, grimacing and grunting as she does so. Mug threatens and lunges at her, and she turns away. Mug continues to stalk Itch. They exchange positions as she evades him. By this time, Mug as well as most of the females are grunting.

At 5:15 Mug abandons his quarry and turns to threaten Pawless and Sol. He moves toward them, then pauses and slaps the ground after them as they move away. At 5:19 Mug turns and races to the tree where Itch and Scratch are. As the mother moves out on a limb, Mug follows with his eyes. Itch's branch snaps, and she falls a short distance. Mug moves toward her at a walk. Itch runs away along a wall with the male following her. Pawless and Sol join Itch, and another chase through the trees ensues. Earlier patterns are repeated until 5:30, when the monkeys are driven away by a gardener.

To summarize events on August 12: Mug repeatedly charged Itch and her infant. On each occasion other females, almost always including Sol and Pawless, two low-ranking females who had no infants of their own that year, intervened. Mug's stalking was single-minded and was sustained over a ninety-minute period. The action was interspersed with long pauses during which Mug may have been dissimulating his intent by staring in another direction while actually maneuvering closer to Itch. By 9:10 the next morning, Mug had already resumed stalking Hillside troop mothers, though on this occasion no contact was made.

During August and September, all four mothers, but especially

Fig. 8.4. Two older females, Sol and Pawless, audaciously intervene to rescue Itch's infant from Mug (6:30 p.m. on September 9, 1972).

Bilgay, went to extreme lengths to avoid Mug. By contrast, the mothers' response to Shifty during this same period was quite different. On September 8, 1972, Shifty chased Mug away and then joined the Hillside troop. He approached Harrieta, who did not move away. She was not protective of her infant and allowed Harry to clamber about the rock where Shifty sat. On September 27, Bilgay allowed Mira to play within inches of Shifty, as did Oedipa with her daughter, Virginia. None of these mothers showed any sign of avoidance or fear.

According to my calculations, both Bilgay's and Itch's infants were sired by Shifty shortly after his take-over in 1971. Given that Mug was probably unrelated to either infant, it is interesting to speculate on his bias for attacking Scratch (that is, in six of nine observed assaults). One possibility is that Mug's antagonism was influenced by Scratch's sex,[4] though more probably Mug was

4. A folk belief that I frequently encountered when asking local people about langurs was that langur males routinely kill all male infants. This theory has the obvious merit of explaining why there were more females than males in troops. However, I found no evidence to support this, and I knew of cases in which both male and female infants were killed by incoming males.

responding to Itch's carelessness. All the Hillside mothers were restrictive of their infants in Mug's presence, but Bilgay was
especially so. She rarely allowed her infant to wander, and she
almost never allowed Mug to get close to her. The close contact may
have been related to Mira's age (she was the youngest infant in the
troop). In contrast to Bilgay, Itch may have been more casual about
allowing Mug to approach. Maternal negligence was almost surely
at issue in the second attack on Scratch that I witnessed on September 5, when Itch let Scratch fall out of a jacaranda tree. Mug was
sitting alertly on a wall fifty feet away. When the infant fell, he
raced to it, reaching it just split seconds before Sol and Pawless,
who had been sitting on the wall on either side of him. The mother
was the last of these four individuals to reach her infant. Only by a
fierce assault were the females able to wrest the infant from his attacker. After Itch had retrieved her infant, Sol persisted in chasing
and slapping at Mug (fig. 8.5).

On August 12 and September 1, Scratch received superficial cuts
as a result of Mug's assaults. On September 9, Scratch was severely
wounded when he was attacked by Mug near the Phiroze. Tooth-
marks were inscribed on his head, and he had a deep gash across
his left thigh (color fig. 7). The crippled infant was helpless without
his mother, and she did not desert him, as has been reported under
similar circumstances at Dharwar (Sugiyama, 1967:229; Yoshiba,
1968:236). When I left Abu on September 12, 1972, I did not expect
that the wounded infant would survive. I was informed later, however, that Scratch did in fact recover and was able to walk again
(Mona Ali, personal communication). Mrs. Ali also told me that
two days after I left, a dark infant was killed by an adult male on the
riding field near Boulders.

When I returned in February 1973, Scratch, Mira, and Virginia
were missing. Based on Mrs. Ali's report, the fact that Boulders
was well within Hillside troop's home range, the conviction that
Mira was the only dark infant in the vicinity at that time, and the
observation that Bilgay was more than five months pregnant in February (which meant that she must have resumed cycling around
September or October), I am relatively certain that the infant killed
by a male was Mira. Based on observed behavior of the two most
likely male suspects, Mug and Shifty, and on the reaction of the
mothers to each of these males, I assume that Mug was the killer.

The cases of Scratch and Virginia are more problematic. In February, both Itch and Oedipa were probably in the first months of
pregnancy, which suggests that either of these infants could have

Fig. 8.5. After Itch had retrieved her infant, old Sol continued to harass Mug for several minutes (5:20 p.m. on September 5, 1972).

been lost as late as December or January. This timing is also in line with Mrs. Ali's report that Scratch had time to recover before he disappeared. The seven observed assaults on Scratch, which included three near misses, make it plausible that Mug was finally responsible for Scratch's death. I suspect that Virginia was also killed, but not necessarily by Mug. On my return, five new males in addition to Mug were present in Hillside troop.

The changes in the Hillside troop as of February 11, 1973, can be summarized as follows: (1) only Harrieta's 15-month-old son had survived from the previous season; (2) around November, Pawless had given birth to a daughter who was 3-4 months old; (3) Bilgay was in the last months of pregnancy; (4) from the extremely mild form of estrous behavior that they exhibited, I inferred that Oedipa and Itch were in the first months of pregnancy; (5) five adult males in addition to Mug were present in the troop whenever he was.

In 1973, females with infants reacted to the five new males just as mothers had reacted to Mug in 1972: by avoiding them. On February 11, the day that I first saw her, Pawless' daughter had superficial cuts on her head in very much the same pattern as the cuts which I had seen Mug inflict on Scratch's skull in September 1972. On February 26 I witnessed males assault Pawless and her infant near Jodhpur House and inflict a four-inch cut down the infant's back. The circumstances surrounding this attack are as follows: On February 26 Pawless and her infant spent the day alone and returned to Hillside troop at 5:20 p.m. No male had been in Hillside troop that day in almost seven hours of observation. Pawless was still with the group at 8:35 the next morning when Righty, followed by five other males, cantered up the road from their sleeping trees beyond the Phiroze School. About thirty minutes later, Kali and other males attacked Pawless and her infant. Even though I watched the February 26 attack from five meters away and I was able to make a Super-8 film record, it was not always clear just which males were attempting to injure the infant. From my firsthand impression and from replays of the film, Kali appeared to be the most persistent attacker. In one sequence, Pawless jumped almost a meter into the air, onto a wall. As she was in midair, Kali leapt up and tried to pull her infant from her chest. Subsequently, Kali pursued Pawless far out onto a limb and forced her to jump to the ground among males who were gathered below.[5]

At the time, I was aware of Itch's attempted, but ineffectual, intervention. The younger female lacked the Boaddicean aggressiveness that both Sol and Pawless had displayed the year before when defending Itch's infant. What I did not notice at the time was the possibility of Mug's assistance. From the film, Mug appears to be supporting rather than harassing Pawless. In one scene, Mug joins Pawless in the center of a cluster of males who have been grabbing at her infant. Even though he is beside her, Mug shows no interest in her infant; his threatening posture is directed outward toward the other males.

After the first of the four attacks, Pawless was panting, diarrheic, and apparently quite shaken. At 9:15, fifteen minutes later, she left the troop for an *Anogueissus* tree one hundred meters away. Minutes later, she descended and sat in the midst of some builders who were quarrying stone, though normally langurs

---

5. Victorian naturalists witnessed a similar strategy employed by an invading male who pursued a mother-infant pair and shook them off the branch that they had escaped to (Hughes, 1884).

avoided such frightfully noisy human activities. When the males, followed by the rest of Hillside troop, headed in her direction, Pawless doubled back in the direction from which she had come. She did not return to the troop that day. When I was able to get a close look at Pawless' daughter (Pawla) seven hours later, I could see that the wound on her back was only superficial. The cut healed within a few days.

It may have been in order to avoid such attacks that Pawless no longer spent much time with the Hillside troop. Of 156 hours that the troop was observed between February 11 and March 25, Pawless and her daughter were absent during 111 of these. Including February 26, the day Pawla was wounded, Pawless and her daughter were present during only eight observation hours when Mug and the five males were with the troop. On ten different days, Pawless and Pawla were spotted traveling apart from the group. On four of these occasions, Harrieta—the only other Hillside mother—and her son were with them. On at least five other occasions, however, Harrieta and Harry were seen with the group when Pawless and Pawla definitely were absent. Pawless' new range was the area of overlap between the Bazaar and Hillside troops' home ranges. She also entered sections which I had previously designated as Bazaar troop's core area, normally used exclusively by them. On March 4, for example, I followed Pawless, Harrieta, and their offspring into the Bazaar, west of Hillside troop's usual range.

After the February 26 attack, Pawless was observed with the troop on five occasions; only once, on March 25, were any of the six males present when she was. Except for this day, Pawless returned to Hillside troop only when Shifty was nearby (a proximity which at this time was invariably correlated with the absence of the six males), or else she arrived in the evening after the males had departed for their sleeping trees to the northeast of Hillside troop's range and she left at sunrise before the males rejoined the group. On at least six occasions Pawless was observed either at the Hillside sleeping trees with the other females or else she was spotted alone on the following morning in the vicinity of their sleeping trees (fig. 8.6).

By staying away from her troop, Pawless was able to keep her daughter alive through March 1973. When I returned in December, Pawla was missing. Hillside troop consisted of the same six females, Harrieta's juvenile son Harry, and a new infant, Bilgay's 10-month-old son, Miro. Mug was the only male ever observed with the troop.

By December of 1973 and January of 1974 Mug no longer ran

Fig. 8.6. After the attack on her infant (February 26, 1972), Pawless and her daughter traveled apart from the rest of Hillside troop for much of the time.

away from Shifty Leftless whenever the older male approached. Although Bazaar troop remained able to displace the smaller Hillside troop on most occasions that the two met at their common border, Mug now remained with his harem as they retreated. On one occasion, Mug even contacted Shifty and grappled with him briefly before retreating. On two other occasions Mug chased away seven males from the Waterhouse band when they attempted to enter the Hillside troop. A newly staunch Mug was now holding his own against outside males.

By 1975 Mug's star had risen. When I returned to Abu in March of that year, Shifty was no longer with the Bazaar troop. In his place was Mug. It was not known what had become of the extraordinary male with the bite out of his left ear. Had he died? (The corpse of an adult male langur was found that spring near Sanand House.) Was he finally usurped by his longtime antagonist, Mug? Did he move on to yet another harem?

Mug's former position in Hillside troop was filled by the young adult male known as Righty Ear. Righty had been one of the five males who invaded Hillside troop in the spring of 1973 and who had subsequently remained in the vicinity of that troop. During January of 1974, months after Righty had entered Hillside troop and passed out of it again, he was seen along with two unidentified males at Eagle's Nest, an estate in Hillside troop's core area. By March of 1975, Righty Ear's waiting game apparently paid off when he came into sole possession of the Hillside troop. But as in the case of his predecessors, Hillside troop was only a stepping stone and Righty Ear aspired to more; by April of 1975 he replaced Mug as the alpha male of Bazaar troop.

The first indication I had of Righty's arrival in Bazaar troop was a report from the inhabitants of Sanand House that an adult male langur had killed an infant at 5 p.m. on April 15, 1975. On the following day when I investigated this report, the young adult male with the unmistakable half-moon out of his right ear was present in Bazaar troop; Mug was nowhere to be found. An elderly langur mother (Short) still carried about the mauled corpse of her infant. By the following day she had abandoned it.

Subsequent to the killing of Short's infant, Righty made more than fifty different assaults on mothers with infants. Despite these attacks, in the weeks that followed, only one other infant disappeared. Five infants in Bazaar troop remained unharmed when my observations terminated on June 20. By the second week of his tenure in Bazaar troop, Righty's hostility was directed toward females generally, as well as toward females with infants. Righty was especially agressive toward Quebrado, an old female without an infant who had intervened in Righty's attacks upon infants. On one occasion after this old female had grappled with Righty, he lunged at her with such force that she was knocked off balance and out of the tree that the two were in.

After Righty switched from Hillside to Bazaar troop, there followed a period of some nine or more weeks when the Hillside females had no resident male except during brief periods when Righty visited them. Whenever the two troops met at their common border, females from Hillside troop approached Righty. Whereas relations between Righty and Bazaar troop females were characterized by extreme antagonism (Bazaar troop females either threatened Righty or avoided him), relations between Righty and Hillside females during the same period could be characterized as "trustful." Righty either sat calmly in close proximity to Hillside females or else engaged in casual grooming with them. The longest time that I ever witnessed an adult male langur groom a female was spent by Righty grooming a very pregnant Bilgay. (Normally males only groom estrous females for any substantial period.) Righty at this time promoted a merger between females in his two harems by solicitous behavior toward Hillside females coupled with fierce chastisement of Bazaar troop females whenever they attacked Bilgay and other Hillside females. Despite the combined efforts of both Righty and Hillside females, the hostility of Bazaar troop females forestalled any merger. By October of 1975 (when Abu was visited by James Malcolm) the two troops were still separate, and the vacuum created by Righty's departure from Hillside troop had been filled by a new male called Slash-neck because of the deep gash in his neck.

Over a period of at least five years, then, political histories of the Hillside and Bazaar troops were intimately linked by a succession of shared usurpers. First Shifty Leftless, then Mug, and finally Righty switched from the small (five to seven females) and apparently rather vulnerable Hillside troop to the larger (eight to twelve females) Bazaar troop (fig. 8.7). It seems possible that the shifts were motivated by the greater number of reproductively active females in Bazaar troop, although the alternative explanations that Bazaar troop's range was more secure from male invaders or superior for foraging cannot be ruled out. It is also possible that each of the three double-usurpers aspired to possession of two troops simultaneously; if this was so, the gamble never paid off, though in the past, or perhaps in the future, some langur male may actually manage this.

Splitear's Ouster from Toad Rock Troop

From June 1971 through March 1973, the composition of Toad Rock troop was relatively stable. Except for the birth of infants, the only recorded changes during this period were the disappearance of one very old female, who was presumed dead, and the disappearance of a nearly adult male and two juvenile-to-subadult males between September 1972 and February 1973. It is possible that these three young males were pushed out by the alpha male.

When recontacted in December 1973, the former alpha, Splitear, eight juvenile and subadult males, and two mothers (Pawlet and Scrapetail) with their partially weaned infants (both approximately 13 months old) were traveling in one body. Occasionally, a third mother (P-M) and her 6-month-old daughter also traveled with them. The remaining five Toad Rock adult females, who did not have infants, were traveling as a separate group, accompanied by a new adult male, Toad. Because almost a year had elapsed since Toad Rock troop was last counted, it was not known how many infants had been born and if any were missing. During January 1974, two additional infants were born. Both Splitear's contingent and this

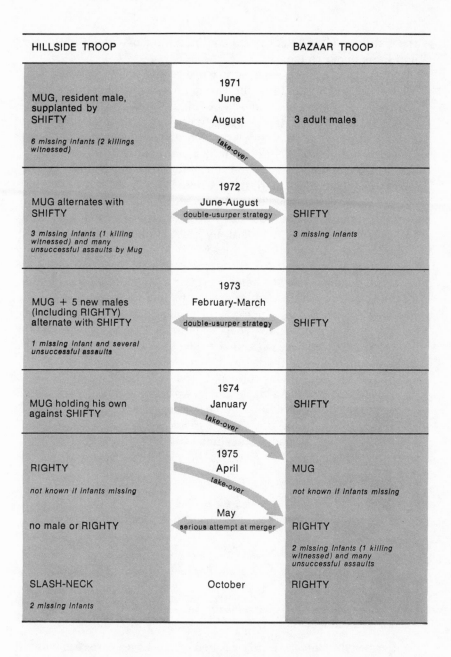

HILLSIDE TROOP		BAZAAR TROOP
MUG, resident male, supplanted by SHIFTY	**1971** June August	3 adult males
6 missing infants (2 killings witnessed)	*take-over* →	
MUG alternates with SHIFTY	**1972** June-August ← double-usurper strategy	SHIFTY
3 missing infants (1 killing witnessed) and many unsuccessful assaults by Mug		*3 missing infants*
MUG + 5 new males (Including RIGHTY) alternate with SHIFTY	**1973** February-March ← double-usurper strategy →	SHIFTY
1 missing infant and several unsuccessful assaults		
MUG holding his own against SHIFTY	**1974** January *take-over*	SHIFTY
RIGHTY	**1975** April *take-over*	MUG
not known if infants missing		*not known if infants missing*
no male or RIGHTY	May ← serious attempt at merger →	RIGHTY
		2 missing infants (1 killing witnessed) and many unsuccessful assaults
SLASH-NECK	October	RIGHTY
2 missing infants		

Fig. 8.7. Summary of the intertwined histories of Hillside and Bazaar troops between 1971 and 1975.

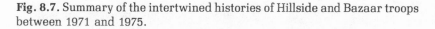

"reconstituted" Toad Rock troop continued to move about in the old Toad Rock home range.

Two weeks after the two groups were recontacted, on January 8, both Pawlet and Scrapetail—the two mothers who had continued to travel with Splitear—were spotted back in the main part of the reconstituted Toad Rock troop, along with their two daughters. On that day, the new male, Toad, attacked and superficially wounded Scrapetail's daughter (N-M). It may have been in order to avoid such attacks on their infants that the mothers had previously stayed away from the reconstituted troop. On January 7, 1974, the day that Pawlet may have returned to the troop, she apparently attempted to leave her daughter (Niza) behind with P-M. It is extremely unlikely that a mother could "accidentally" abandon her 13-month old infant; Pawlet's desertion was apparently intentional. Furthermore, the "maternal concern" for the safety of her infant exhibited by Pawlet on the next day (see below) argues against negligence in this instance. In any event, Pawlet's attempt to separate herself from Niza was unsuccessful, since all three mothers and three infants had returned to Toad's new troop by January 8. Possibly the female that Niza had been left with also returned. Other alternatives include the possibility that Niza independently made her way back to her mother or that Pawlet returned for her daughter.

Events leading to Niza's desertion as well as the infant's response to it are described below for the period between December 21 and January 8, 1974. On the days that Pawlet and Scrapetail were not with either group, it must be assumed that they were traveling on their own as Harrieta and Pawless did in February and March of 1973.

December 21. Pawlet and Niza were spotted with Splitear and the eight young males from Toad Rock troop near Nakhi Lake, foraging beside the ashram.

December 23. Splitear's band of males meets the School troop and pushes them into their core area. Scrapetail, Pawlet, and their infants are with them.

December 24. Splitear's band meets the School troop near Windermere Lake. It is not known if any females are with them.

December 28. Splitear's band is spotted but no females are seen with them.

December 29. D. B. Hrdy observes encounter between Splitear's band and School troop near Windermere Lake. It is not known if the females are with them.

From 9:10 to 3:30 the reconstituted Toad Rock troop forages near Jaipur quarters, moving gradually toward the diary. They are so spread out that I observe them for over four hours before I feel I have a complete count; neither Pawlet, Scrapetail, nor P-M are with them.

January 1, 1974. Splitear's band forages in proximity to School troop next to Shiv Kuti. Pawlet, Scrapetail, P-M, and their three daughters are with the band. On this day, subadult males from Splitear's band raid School troop and steal copulations there.

January 6. Around 1:00 p.m., Splitear's band, including at least two of the mothers and their offspring, encounters the newly reconstituted Toad Rock troop near Jaipur House. The ensuing hostilities involve teeth-grinding and chases. Because of the dense vegetation in the canyon below Jaipur House where much of the action occurred, no definite identification of the participants was made. Identification of unseen combatants was complicated by the fact that the Waterhouse band of seven males was also in the neighborhood of Toad Rock troop at this time, and was definitely involved in at least some of the chases.

After this encounter, members of Splitear's band were in disarray. At 7:42 the next morning I encounter a lone juvenile male still wandering separated from the band. At five o'clock on the afternoon of the January 6 encounter, Pawlet and her daughter Niza were seen traveling with the new male and the main body of Toad Rock females.

January 7. At 7:30 Pawlet and P-M and their infants are in a fig tree beside my bungalow. The four remain there until 8:44 when Pawlet strikes out by herself toward the *Anogueissus* grove above Jaipur quarters, which is a regular sleeping site for Toad Rock troop. Moving quickly, before her daughter can follow, Pawlet leaves Niza behind with P-M. At 8:55 I hear whoops from the direction in which Pawlet headed. At 9:08 there are three spaced whoops from the *Anogueissus* grove. I am relatively certain that Toad and the Toad Rock females are there.

From 9:30 on Niza gives high-pitched birdlike calls, which alternate with a whimpering noise unlike any I had ever heard from a langur. At 10:05 P-M heads in the opposite direction from the *Anogueissus* grove; Niza follows her. They cross into the Toad Rock-School overlap, with Niza continuing to emit cries. Niza climbs high into an *Anogueissus* tree where she alternates between giving cries from a high lookout post and coming down into the tree to play briefly with P-M's 6 month-old daughter.

At 10:20 Niza ignores play solicitations from the other infant. Perched at the top of the tree, she gives a series of high-pitched cries at about six-second intervals. At 10:28 P-M and her infant move further uphill, but Niza remains at her post. At 10:32 Niza stands bipedally on a branch for a better view. She calls, "Mek, mek, mew," and stares off toward the bungalow. At 10:40 a young male from Split-ear's band appears. I lose Niza in the rugged terrain, but locate her again within minutes by her cries.

By 10:55 Niza has rejoined P-M and plays with her infant. At 11:00 P-M heads southward, looking around her with circumspection. At 11:10 Niza resumes her calling. Over a two-minute period, she calls thirteen times. At 11:47 one of the subadult males from Splitear's band approaches her. Niza gives an alarm chirp and climbs higher into the tree she is in. Splitear and seven young males appear and forage in the vicinity of P-M, P-M's daughter, and Niza. In the course of foraging, the male band drifts away from the three females; I follow the males.

At 3:17 I contact the reconstituted Toad Rock troop near the dairy behind Jaipur House. Pawlet is with them, and so is Niza. On one occasion, Pawlet briefly allowed Niza to suckle before moving away. P-M, Scrapetail, and their two daughters are also with Toad's troop. January 8. At 8:40 the reconstituted Toad Rock troop is foraging in the *Anogueissus* grove above Jaipur quarters. Both P-M and Pawlet have fierce weaning spats with their daughters. Pawlet resolves the dispute by making a leap between trees which is greater than Niza can negotiate.

At 10:55 the new male, Toad, attacks Scrapetail's daughter, N-M, and wounds her superficially on her left arm and shoulder. At 11:44 shortly after this attack, Niza gives a little cry (cause unknown). Pawlet rushes to her daughter and gathers her up.

At 12:50 the wounded infant solicits Toad by presenting from ten meters away and frantically shaking her head. Subsequently, N-M approaches Scrapetail and is allowed by her mother to suckle briefly.

After January 8 none of the mothers or their infants were seen apart from the main body of the troop. No further attacks by Toad on infants were seen, though all mothers with infants made obvious efforts to avoid him within the troop. Toad's movements or his approach elicited grunts and occasionally squeals from the mothers; they would gather their infants to them and move some distance away.

Scrapetail's 13-month-old daughter recovered within days from her superficial wounds and a slight limp, though she retained a scraggly appearance and seemed to be in poor health. N-M's malaise may have been related to the abruptness of her weaning, but unfortunately, detailed comparative data on this point were not recorded. After Toad's take-over, all three Toad Rock mothers were making active efforts to discourage their infants, aged 6-13 months, from suckling. Suckling was still permitted in each case, but usually ended in fierce weaning squabbles which the infant lost.

Within weeks of their return, all three mothers exhibited estrous behavior and solicited Toad. On January 11, 12, 13, 17, 19, 21, and 26 Scrapetail solicited the new male; on January 30 she had traces of menstrual blood about her perineum. Pawlet came into estrus from January 21 to 23. P-M was observed soliciting the new male on January 22. For days after the January 8 attack on N-M, she, too, addressed estrous signals to Toad.

Fission in the I.P.S. Troop

Between 1971 and 1975, the I.P.S. troop grew rapidly, almost doubling in size from 24 to 40 individuals. By 1975, however, the former leader (LeGrand), all immature males, and all infants had disappeared. Females from the former I.P.S. troop, with their newborn infants, and juvenile and subadult females were divided between two new males. Both harems remained in the same area as the former I.P.S. troop but for reasons unknown to me had become extremely shy of humans and difficult to study. One group was composed of a single adult male, seven adult females, one subadult female, three dark and two cream-colored infants. An exact count was never obtained for the second group. Nor was it known how the former home range was apportioned between them or if the boundaries had changed in any way.

Population Density, Political Change, and Infanticide

Fourteen take-overs from outside the troop and at least one semi-permanent invasion by a male band have been reported at Dharwar, Jodhpur, and Abu. At least eleven of these were accompanied by witnessed assaults on infants or else by the disappearance of unweaned infants.[6] To date, the entrance of new males into troops has coincided with the recorded death or disappearance of some 39 infants (table 8.3). Assaults by langur males upon infants have only been reported when a male entered (or reentered) the troop from outside it.

A definite correlation exists between political change and infanticide. A more tentative correlation may also exist between the fre-

---

6. Male invasions at Chippaberi were also accompanied by the wounding, death, or disappearance of unweaned infants. Because of the intermittent nature of my observations on this troop, circumstances surrounding these invasions are unknown.

TABLE 8.3. Political changes and infanticide at Dharwar, Jodhpur, and Mount Abu.

Location and date when change was noted	Troop	Source	Invading males	Resident adult males		Males excluded		Infants killed or missing
				Before take-over	After take-over	Invaders	Residents	
Dharwar (June 1962)	30th	Sugiyama, 1965b	7	1	1	6	1 adult, 6 juveniles	6
Dharwar (July 1962)	2nd	Sugiyama, 1966	1	0	1	0	0	4
Dharwar (July 1962)	5th	Yoshiba, in Sugiyama, 1967	59 (including males X, Y, Z)	1	1	58 (ca.)	1	?
Dharwar (Oct. 1962)	7th	Yoshiba, in Sugiyama, 1967	3 (X, Y, Z)	1	1 (X)	2 (Y, Z)	1	?
Dharwar (March 1963)	1st	Kawamura in Sugiyama, 1967	2 (Y, Z)	3	2 (Y, Z)	0	3	4-5
Jodhpur (July 1969)	B-26	Mohnot, 1971b	22	0	1	21	0	5
Mt. Abu (July-Aug. 1971)	Hillside	This study	1a (Shifty)	1 (Mug)	1 (Shifty)	0	1 adult (Mug) 1 juvenile	6

Mt. Abu (May 1972)	Bazaar	This study	1 (Shifty)	3	1 (Shifty)	0	3 adults, 2 juveniles	4
Mt. Abu (June–Aug. 1972)	Hillside	This study	1 (Shifty)	1 (Mug remains)	1 or 0	1	1	3 (Scratch, Mira, and Virginia between Sept. and Feb. '73)
Mt. Abu (Feb. 1973)	Hillside	This study	5	1 (Mug remains)	6 or 0	5 or 0	1 or 0	1 (Pawla)
Mt. Abu (Dec. 1973)	Toad Rock	This study	8^b	1 (Splitear)	1 (Toad)	7 ?	1 adult, 8 subadults	?
Mt. Abu (April 1975)	Bazaar	This study	1 ?	1 (Shifty)	1 (Mug)	1 ?	1	2 juvenile females
I.P.S.			2 ?	1 ?	Troop splits	?	?	?
Mt. Abu (May 1975)	Bazaar	This study	1 (Righty)	1 (Mug)	1 (Righty)	0	1 ?	2
Mt. Abu (Oct. 1975)	Hillside	This study	1 (Slash-neck)	0	1	0	0	2
		Total: ca. 114				ca. 133		39 (minimum)

a. The actual number of invaders is not known, though I never had reason to suspect that Shifty acted in concert with a band.

b. The actual number of invaders is not known, though I suspected that Toad was formerly a member of the Waterhouse band. This suspicion was based on the extraordinary physical resemblance between Toad and two other males in the Waterhouse band, suggesting that they might be brothers, and the repeated attacks made by this band after Toad took over, even though previously they had never been observed in the Toad Rock range.

quency of political changes and population density. At Orcha and Kaukori where langur population densities are low (three to six langurs per square kilometer), only one instance of change was observed. This shift involved the replacement of a leader by another troop member and took place with relatively little conflict (Jay, 1965). Nor have take-overs been reported by either of two field teams working in the sparsely populated Himalayan districts of Nepal (Bishop, 1975; Curtin, 1975). However, each of these studies was comparatively short (4 to 17 months) in duration. Are take-overs rare at low densities? Do they occur at all? Or did they simply not occur within the short span of these studies?[7] Only long-term studies of langurs in sparsely populated areas can provide answers to these questions.

All long-term studies of langurs have been in areas with much higher population densities. At Dharwar where population densities range between 84 and 133 langurs per square kilometer (Yoshiba, 1968), males from outside the troop wrested control of four of nine troops under surveillance between March 1961 and April 1963 (Sugiyama, 1967). Including surveyed troops, ten major changes, six witnessed and four inferred, were reported in a two-year period (Yoshiba, 1968). At Abu, where population densities exceed 49 animals per square kilometer, nine major changes occurred in four of five troops over a period of 252 troop-months—an average of 0.43 take-overs per year.[8] Data on population density and on rates of change have not yet been published for Jodhpur, but it is worth noting that because of the extremely arid conditions under which these langurs live, they tend to concentrate about intensively used garden spots and watering places, even though vast open spaces are available to them. Similar correlations between adult male replacements, death of infants, and high population densities have been reported for *Presbytis senex* in Sri Lanka (Rudran, 1973b).

7. No take-overs were reported for Naomi Bishop's field site at Melemchi, but during the last months of her study, in the monsoon mating season, one adult male left her main (multimale) study troop and a new male joined it. Additionally, sightings of extratroop males and the incidence of woundings showed a marked increase during this period (Bishop, 1975b). Similarly at Solu Khumbu, Curtin (1975) reported shifts in adult male membership of the troop during the breeding season. These observations suggest that even at low densities the breeding season is not completely competition-free.

8. This rate can only be a minimum estimate, but I have no reason to suspect that it does not accurately reflect the true incidence of take-overs in the I.P.S. and Toad Rock troops between June 1971 and June 1975 and in the Bazaar, Hillside, and School troops during the period between June 1971 and October 1975, when J. Malcolm visited these three troops.

As Yoshiba (1968) and others have pointed out, the most obvious link between population density and the frequency of take-overs is the increased numbers of extratroop males at high densities. If the possibilities for male recruitment are greater at high densities, and if a band of males has a better chance of successfully usurping a bisexual troop than would a single male, then there will be more take-overs in crowded areas. It is also possible that the population of loose males will be augmented by take-overs. As can be calculated from table 8.3, the number of adult and juvenile males expelled from troops will on average exceed the number who actually invaded it. This increase in potential invaders may be aggravated by a ricochet effect, as seen in Dharwar's 5th, 7th, and 1st troops (table 8.3). Within a matter of months, males X, Y, and Z, who were evicted from one troop, jointly attacked another, followed by ousted males Y and Z together usurping yet a third troop (Sugiyama, 1967).

An alternative explanation for the correlation with high densities has been offered by Rudran (1973b). According to him, take-overs occur in order to maintain the one-male troop structure and infanticide occurs so as to curtail population growth in crowded areas. Unquestionably, one-male troops and reduced infant survival are outcomes of the take-over pattern. But if take-overs and infanticide are advantageous to the individual males who practice them, then these outcomes are only secondary effects—not explanations for observed behavior.

Despite the apparent correlation between density and the rate of political changes and the existence of plausible explanations for it, chance, historical factors, and the individual personalities of the males involved are clearly important. In two cases, for example, take-overs occurred only after the absence of the resident troop leader was discovered by neighboring males. At Dharwar, the leader of the 2nd troop was experimentally removed before the leader of neighboring 4th troop entered and killed all infants (Sugiyama, 1966). At Jodhpur, the B-26 troop was rendered vulnerable to attack by a male band when the leader and 68 troop members died, apparently from drinking contaminated water (Mohnot, 1971b). At Abu, nomadic extratroop males were able to enter the small and peculiarly vulnerable Hillside troop for four years in a row. Over a 38-month period between 1971 and 1974, ten of twelve infants born in Hillside troop—that is, 83 percent—disappeared. Almost concurrently, beginning in 1972, Shifty took over the Bazaar troop, killed some four infants, but then remained with the troop for a period of three years, allowing no other males to enter. Whereas Hillside troop during this period faced the possibility of extinction, the

Bazaar troop was characterized by stability and growth. In this instance, a fortuitous affiliation with one troop rather than another apparently determined a female's fate.

It has been suggested that the high population densities reported for sites like Dharwar are recent, the result of deforestation and compression of the langur's habitat as well as reduced predation (Sugiyama, 1967; Crook, 1971). One of the implications of the view that langurs are living under new conditions to which they are not adapted is that the aggressive behavior reported for crowded langurs is somehow pathological, that it represents "destruction—not adaptation" (Dolhinow, 1977). This is the view which prevailed at the outset of my study, and one which persists today (Curtin, 1977). Six different lines of evidence, however, argue against the social-pathology hypothesis. The accounts of early naturalists, dating from the nineteenth century, indicate that intense competition between langur males for females, and resulting high levels of aggression, have characterized this species for well over one hundred years, and probably much longer. Additional evidence for the antiquity of the pattern of male take-overs and infanticide derives from apparently shared predispositions among widely divergent members of the subfamily Colobinae. In contrast to pathological behavior, observed assaults upon infants have been highly goal-directed, specifically aimed at infants and rarely injuring any other party. Furthermore, infanticide has only been reported when males enter the breeding system from outside it, and the direction of assaults upon infants has been consistent with probable paternity; that is, the infants attacked were almost invariably those sired by competitors. The existence of a number of female counter-strategies against infanticidal males suggests that infanticide has impinged on females long enough for such traits (to the extent that they are genetic) to have evolved. Finally, and most importantly, the available evidence indicates that males do gain from killing infants.

The Evolution of Infanticide

Sugiyama (1966) estimated that at Dharwar political changes take place once every three to five years. In a later publication (1967), he allowed for an even faster rate of change, once every 27 months, a figure which is astonishingly close to the 27.6 month average tenure for langur males at Mount Abu (the basis for this calculation is provided in appendix 4). Such rapid rates of male take-overs place great pressure on a langur male to compress as much as possible of his harem's reproductive output into his brief tenure.

If conditions such as those at Dharwar and Abu are in fact not due to recent changes, then intense competition between males for females, a fast rate of political change, and resulting pressure on males to maximize their reproductive success within a short period constitute the conditions under which some langurs evolved. By eliminating infants in the troop that are unlikely to be his own, a usurping male hastens the mother's return to sexual receptivity and reduces the time that will elapse before she bears his offspring. Infanticide then might permit an incoming male to use his short reign more efficiently than if he allowed unweaned infants present in the troop at his entrance to survive. This hypothesis assumes that in competition between males for females, taking over a troop is more or less equivalent to reproductive access to females and that insofar as these females are fertile, such access will be correlated with differential reproduction. To the extent that death of her infant induces a female to ovulate in the presence of the male who killed her infant, infanticide should on average increase the reproductive fitness of those males who practice it.

Once infant killing began, a usurper would be penalized for *not* committing infanticide. If this male failed to kill infants upon taking over the troop, and instead waited for those infants already in the troop to be weaned before he inseminated their mothers, then his infants would still be unweaned and hence vulnerable when the next, presumably infanticidal usurper entered. The advantages to an infanticidal male migrating to a population in which noninfanticidal males predominated should be obvious. Nevertheless, a recent computer simulation, of the langur breeding system by Chapman and Hausfater (in preparation) raises the issue of tenure lengths at which infanticide would *not* be advantageous. At a 36-month tenure length, for example, infanticidal males have lower reproductive success than their noninfanticidal counterparts, according to the simulation, because females would still be pregnant with a noninfanticidal usurper's second crop of offspring when an infanticidal usurper entered, and infants born after a take-over are not attacked —even though they may have been sired by the previous male.

Other fluctuations in tenure length might likewise be expected to select for changes in male behavior. For example, if the rate of take-overs was to be speeded up and then held constant, male tolerance towards immatures might be drastically altered. With a faster rate of take-overs, it would be unlikely that the same male could remain in control of the troop long enough for immature females to reach menarche and give birth to a new infant that would in turn grow old enough to survive the next take-over. Immature females would be worth no more to the usurper than young males, and might compete with the productive members of his harem for resources. Under these circumstances, it would behoove a usurper to drive out immatures of both sexes. This is precisely what occurs among *Presbytis senex* living at very high densities (as high as 215 per square kilometer) at Horton Plains in Sri Lanka. The ousted females travel with former male troopmates in predominantly male, mixed-sex bands (Rudran, 1973b).[9] From this point of view, Toad's attack on an almost weaned female (13-month-old N-M in 1974) was a short-sighted strategy. Whether subsequent attacks on N-M were forestalled by her age and nearly weaned status or by her sexual solicitation of Toad is not known.

Precise measures of reproductive success for males who take over a troop and kill infants as well as for those males who do not are crucial to the acceptance of the sexual selection hypothesis. Unfortunately, even though paternity exclusions from primate blood samples are technically feasible, for political reasons, our plans to gather such samples at Abu had to be abandoned. Nevertheless, in at least three troops (the 30th troop and the experimental 2nd troop at Dharwar and the Bazaar troop at Abu after Shifty's take-over in 1972), females whose infants were killed were observed copulating with the new male, and 70 percent of these females (four of six in 30th troop; three of four in 2nd troop; and four of five in Bazaar troop) gave birth within six to eight months, just over one langur gestation period. Barring the possibility that another male besides the usurper was actually the progenitor, it is hard to see how infanticide in these cases failed to increase the reproductive success of those males who practiced it.

In this context, it can not be stressed enough that no conscious intent on the part of the male is implied. It is simply assumed that the genes of animals who respond to a given situation in the most advantageous way—relative to other animals in the population—will be disproportionately represented in the next generation. In order

9. Based on intermittent observations at Horton Plains, Rudran estimated that take-overs occurred once every 36 months ± 23.1. The fact that immature females as well as males are excluded suggests that actual tenure lengths may tend toward the short end of this spectrum (that is, closer to one year). Alternate explanations for the ouster of immature females among *P. senex* include: age of first pregnancy may be delayed among the *P. senex* of Horton Plains; intratroop competition for resources might be greater at Horton Plains; or a male might be more likely to be succeeded by a close relative who would benefit from a large harem among *P. entellus*.

for an infanticidal male to leave a disproportionately great number of offspring, however, he must possess some means of descriminating between infants likely to be his own and those belonging to another male. A male who attacked his own offspring would be rapidly selected against. To some extent, differential treatment of infants is incumbent in male roles (Ransom and Ransom, 1971; Blaffer Hrdy, 1976). Among langurs, for example, a troop leader (presumably the progenitor of infants in his own troop) is normally tolerant of infants, active in troop defense, and occasionally heroic in the rescue of infants. Troop leaders have a vested interest in keeping other troops and especially extratroop males away from their harem and offspring, and behave accordingly. By contrast, lower-ranking adult and subadult males within the troop are less likely to be progenitors and also to participate in troop defense.

The behavior of extratroop males is quite different. These nomadic males haunt the vicinity of bisexual troops. If they do enter, they ignore infants except when their intentions are specifically hostile. One exception to this rule, however, has been reported in which an invading male held a newborn infant on his lap (Yoshiba, 1968). Unfortunately, no further information regarding the relationship of this particular pair is available. Generally, alien males are either scrupulously avoided or else attacked by mothers with and without infants, though estrous females may approach them.

Even beyond differential treatment of infants by males fulfilling different roles, some males behaved in accordance with a roughly accurate, presumably situation-determined ability to discern probable paternity (see Sugiyama, 1966)—that is, the behavior of some males coincided with information (admittedly incomplete) recorded by human observers. According to my calculations at Abu, for example, the two infants born in Hillside troop in March and April of 1972 must have been conceived around the time of observed copulations in August and September of 1971, just shortly after Shifty took over the troop. Whereas Shifty was tolerant of these two infants, whenever the pre-August 1971 troop leader Mug was with the troop, he repeatedly attacked, and, in the case of Scratch, several times wounded these infants which were in all probability not his. Pawless' infant Pawla was conceived around April of 1972, just before Shifty left the Hillside troop for Bazaar troop. Even though Pawla was born around November, while Shifty was no longer in sole possession of that troop, whenever he visited, he tolerated Pawla just as he would an infant born in his own troop. The mother made no attempt to avoid Shifty, and, in fact, throughout February and March of 1973 she sought him out. Had the resident females in Bazaar troop permitted it, I suspect that Pawless would have joined Shifty's new troop.

It would be of great interest to know exactly what means the langur male has at his disposal for discriminating his own infants from those of a previous leader. One possibility is that his behavior is influenced by previous consort relationships with the mother (as reported for anubis baboons in Ransom and Ransom, 1971), or the lack of them (as in the case of the crab-eating macaques). Familiarity alone, however, does not explain Mug's returning to Hillside troop a year after his ouster as alpha male and attacking infants which almost surely were sired by his rival, Shifty. Barring coincidence, one possibility is that males are roughly able to evaluate the timing of previous consort relationships. If this is so, the mechanism may be imperfect. There is one problematic account of a langur who attacked an infant which probably was his own (S.M. Mohnot, personal communication). More often, however, the converse has been observed: a troop alpha allows another male's offspring, born after his take-over, to survive. At Dharwar, for example, the alpha male of 4th troop was able to unite females from the neighboring 2nd troop with his own harem. As sexual activity in 2nd troop subsided, he returned to his own troop, whereupon another male from the adjacent 3rd troop began consorting with the leaderless 2nd troop. Four months after the male from 3rd troop's accession to 2nd troop, four infants, sired by the first usurper (the alpha male from 4th troop), were born. Subsequently, 2nd troop merged with 3rd troop, yet the 3rd troop male never attacked the new infants who were not his own (Sugiyama, 1966). The fact that females in 2nd troop exhibited estrous behavior and solicited the new male, even though they were already pregnant, may have been an important inducement for the tolerance shown by 3rd troop's alpha male.

Besides the absence of precise information concerning male reproductive success, there are several other problems that complicate confirmation of the sexual selection hypothesis. Though infanticide is a relatively effective means of inducing estrous behavior (within about eight days of an infant's death; see table 8.4), in many cases it is an imperfect means of inducing a female to ovulate, conceive, and to subsequently give birth to a live infant. In the case of female Ti at Jodhpur, for example, 27 months elapsed between her infant's death and the birth of a subsequent one. At Jodhpur the average time between the killing of an infant and the next live birth was 17 months, 2.5 times the average for Dharwar and Abu. One possibility is that this lag may be imposed by the harsh desert conditions at Jodhpur. In a study of reproductive cycles among the related species, *Presbytis senex*, Rudran found that the birth interval at his Polonnaruwa study site was six to nine months longer than at Horton Plains, where more food was available year-round (1973a).

Whatever its cause, however, the cost of this lag is borne largely by the female whose infant was killed, not by the male who killed it. Assuming that he is the father of her subsequent offspring, however delayed, the incoming male may fare better than if he had waited for the mother to wean her first infant and resume cycling naturally. By that time the usurper might no longer be in control. Another risk for the male is that after inducing a female to ovulate, he may not be able to retain exclusive sexual access to her. At both Jodhpur and Abu, females who resumed estrus after the death of their infants solicited males outside the troop. Even if a usurping male is successful in inseminating females, if he then leaves the troop, another male may come in after the infants are born and kill them—as in the case of the ambitious Shifty after he left Hillside for Bazaar troop. Nevertheless, if Shifty's switch was a gamble, he had good odds, since Bazaar troop offered twelve females of breeding age, compared to only five in Hillside troop, and since one or more of the infants he sired in Hillside troop just might make it to weaning.

Another element that may contribute to less than the maximum possible reproductive success for infanticidal males is incompetence. Shifty, for example, might have been a more proficient infanticide than either Mug or Righty, who attacked the same infants again and again without managing to kill them. To what extent infanticide is learned behavior is unknown; it seems unlikely to me, however, that adequate opportunity for such learning is present.

Female Counter-strategies

In almost every instance that the strategy of infanticide seems inefficient, the cause can be attributed either to interference from another male or to noncooperation from females. Confronted with a population of males who are competing among themselves, often with adverse consequences for females and their offspring, natural selection would have favored those females inclined and best able to protect their interests.

When an alien male approaches a troop, he is chased and harassed by females as well as by the resident male. Even when the resident male, for whatever reason, tolerates alien males, females with infants or juveniles may not; they may either actively resist

TABLE 8.4. Approximate time elapsed between death of infant, estrous behavior, and birth of subsequent offspring in three cases of infanticide.

Troop and location	Female	Date of infant's death	Estrous behavior observed	Days elapsed between death and estrus	Date of subsequent birth	Months elapsed between observed estrus and next birth
30th troop	A	June 6-11, 1962	June 12-16	1-5	Not observed	> 9
Dharwar (Sugiyama, 1965b)	B	June 6	June 12-13	6	Late December to mid-January, 1963	Ca. 7
	C[a]	June 29	Not observed	—	Late January to early February	(Ca. 7.5?)
	F	June 17-18	June 23-25	Ca. 6	Late January to mid-February	Ca. 7.5
	J[b]	July 9	July 18	10	Early to late January	Ca. 6.5
	E[c]	August 4	September 6	Ca. 33	Late December to mid-January	4.5

				Female disappeared	
2nd troop Dharwar (Sugiyama, 1966)					
K	June 28, 1962	July 11	14	Female disappeared	
R	June 30	July 15	15	Mid-January to February 5, 1963	6-6.5
S	June 29	July 11-18	13	Mid-January to February 5	6-6.5
U	June 28	June 29	1	Mid-January to February 5	6.5-7
B-26 troop Jodhpur (Mohnot, 1971b, and personal communication, 1973)					
Ti	July 24, 1969	July 28-31	5	October 1971	27
Ni	July 27	August 5-8	9	June 28, 1970	11
Ri	August 3	August 4-7	1	August 1970	12
Pi	July 29	August 8-12	11	(a) Stillbirth (b) July 1972	19
	Average (deleting female E)		8		

a. Presumably, female C came into estrus about the time of the take-over and the death of her infant (June 1962), since she gave birth 7.5 months later.

b. There is a discrepancy in Sugiyama (1965b) between table 3 and the text (p. 398); it is assumed that the date mentioned in the text and cited here is the correct one.

c. Female E's infant was more than a year old at the time of the take-over, and it is possible (note short gestation period) that she was already pregnant, and that her estrous behavior was mock.

such males or leave the troop. After a new male takes over, females may form temporary alliances to prevent him from killing their infants (for example, Sol and Pawless' combined front against the infanticidal Mug).

Females are often able to delay infanticide; less often are they able to prevent it. For this reason, one of the most effective counter-infanticide tactics may be postconception estrous behavior. That is, if males are actually able to evaluate past consort relationships, a pregnant female may induce a male to tolerate her subsequent offspring (not necessarily his) by soliciting this male in the months before her infant if born.   Though the term "estrus" usually connotes a cycling female, postconception periods of pseudoestrus have been reported for a number of primate species, including rhesus macaques (Conaway and Koford, 1965), patas (Loy, 1974), langurs (Ripley, 1965), Japanese macaques (Hanby et al., 1971), chimpanzees (van Lawick-Goodall, 1969), as well as captive gorillas (Hess, 1973), Sykes monkeys, and vervets (Rowell, 1972a). Interestingly, infanticide has been reported in four of these eight species, and it is predicted to occur in a fifth (patas monkeys). If, in fact, postconception estrus is a strategy to forestall infanticidal males, it does not explain why pseudoestrous solicitation should continue into pregnancy in species where infanticide does not occur. For example, no cases of infanticide have ever been reported for vervets, Sykes monkeys, or Japanese macaques. Given the very short lactation interval in vervets (Rowell, 1972a), the apparently "permanent" state of estrus among vervet females (Rowell, 1972b), and the pronounced birth seasons of Japanese macaques (Itani, 1959), one would not expect infanticide to be a particularly advantageous strategy for males in these species. Very little field information is available for the social behavior of Sykes monkeys (Aldrich-Blake, 1970).

At Dharwar, Jodhpur, and Abu, pregnant females confronted with a usurper displayed the traditional estrous signals—present-

---

10. Deceit, of course, is the classical antidote to infanticide, calling to mind the plight of the Greek Titaness Rhea. As each of her offspring emerges from the womb, they are swallowed by Cronos. But as Zeus is about to be born, Rhea devises a plan so that the birth of her baby will be concealed. When the infanticide arrives, she hands him an enormous stone wrapped in swaddling clothes. Cronos of course thrusts the stone into his belly, "Wretch! he knew not in his heart that in place of the stone [Zeus] was left behind, unconquered and untroubled, and that he was soon to overcome [Cronos] by force and might and drive him from his honours, himself to reign over the deathless gods" (Hesiod, ca. A.D. 720, from a translation of the Theogony by H. G. Evelyn-White).

ing to him and shaking their heads. The shuddering in such pseudo-estrus is usually less frenzied than normal. Female E in the 30th troop at Dharwar exhibited estrous behavior and copulated with the new leader even though she gave birth four months later, three to four months short of a normal gestation period. Females in the 2nd troop at Dharwar exhibited estrous behavior fewer than four months before delivery (Sugiyama, 1965b; 1966). Female O at Jodhpur exhibited estrous behavior in early August and gave birth in January, five months later (Mohnot, 1971). At Abu, Pawless exhibited estrous behavior in early August and again in early September and was seen at different times in consort with both Shifty and Mug, even though she was four to five months pregnant with her daughter Pawla at that time. Pawla was probably conceived in March of 1972, while Shifty was still in sole control of the troop.

The presence of male invaders may stimulate a female to exhibit estrous behavior even if (or perhaps especially if) she is pregnant. In a study of corralled patas monkeys on La Parguera Island, Loy (1974) found that the frequency of postconception periods of estrus was higher just after a new male (who quickly became the new alpha male) was introduced. In the presence of this unfamiliar male, there was a mean of 2.8 postconception periods of estrus for five pregnancies; four of these five females had been inseminated by the previous alpha male. In seven other pregnancies that occurred after the new male was well established and the old alpha removed, Loy found that the average number of postconception periods of estrus went down to 0.7 periods per pregnancy on average; the new alpha was the father in all seven cases. Whether these females are actually confusing the issue of paternity awaits further investigation.

After birth an infant's survival is best insured if its mother is able to associate with the father, or at least with the male who considers himself the father or who acts like one: that is, a male who tolerates her infant. Such a strategy might explain Pawless' shift in allegiance between August 1971, when she fought Shifty, and March 1973, when she sought him out. Similarly, the three mothers from Toad Rock troop, who preferred to travel with Splitear's band of ousted males, may have felt safer in the company of the probable father of their infants.

If none of the above tactics is feasible, a mother may leave the troop, as Pawless and Harrieta did in February of 1973 and as Pawlet and Scrapetail did in December of that year (see also the case of female K at Dharwar, listed in table 8.4). Leaving the troop, however, may not be an equally attractive option for all females. Whereas old females frequently left the troop for periods of several hours or more, young females were never observed to do so. If departure is not feasible, and if the infant is attacked and wounded, a mother may continue to care for it or abandon it. Normally a langur mother will carry her infant for hours or even days after it dies; I observed this on three occasions. Yet, in a number of cases of infanticide reported for Dharwar and Jodhpur, mothers abandoned their murdered infants soon after or even before their death (Sugiyama, 1967; Mohnot, 1971b). Females Ti and Ni at Jodhpur held their dying infants less than 25 minutes before abandoning them to scavenging birds.

Rudran has suggested that the mother abandons her wounded infant for fear of injury to herself and "because an adult female is presumably more valuable than an infant to the troop" (1973b). It is far more likely, however, that desertion reflects a practical evaluation of what *this* infant's chances are, weighed against the probability that her next infant will survive. It was a waste of energy for Itch to continue to care for her wounded infant in 1972; soon after the infant recovered, he died (presumably finally killed by Mug). Given the circumstances, desertion and a quick return to estrus might have been the mother's optimum strategy.

Under some conditions, a mother may choose to desert an unwounded infant, although this tactic is preferable to infanticide only if the infant has some chance of survival on its own. In January 1974 Pawlet, who had been traveling apart from the newly usurped Toad Rock troop, attempted to leave her partially weaned 13-month-old daughter in the company of another mother and return alone to the main body of the troop.

The high degree of relatedness among females in a troop almost certainly underlies the cooperation among females in the face of infanticidal males and the extraordinary altruism exhibited by females in the defense of other females' offspring. But in addition to inclusive fitness, reciprocity might be at issue. One year after Itch's infant was defended by Pawless, Itch was the only female to come to the rescue of Pawless' infant when she was attacked by male invaders in Hillside troop. Explanations based on inclusive fitness and reciprocity are by no means incompatible. In fact, the permanent community of long-lived langur females provides precisely the sort of situation in which genes for altruistic females might be selected for (Trivers, 1971). The outstanding question is why these females fare so poorly in the face of males.

Despite the various tactics that females may employ to counter infanticidal males, when a mother is confronted with a male determined to kill her infant, the odds are in his favor. But females do have one last recourse. To the extent that infanticide is advantageous for males, and to the extent that it is a heritable disposition, a female may "choose" to breed with an infanticide so that her own sons will profit from killing another male's offspring. And herein lies the weakness in the combined female front.

Infant Counter-strategies

There is little that a very young infant can do to prevent the attacks of an infanticidal male. Clinging to the mother may be the extent of its options. Among older infants approaching an age where weaning becomes feasible, the separation process may be speeded up after a take-over. It appears that the responsibility for this early termination lies with the mother (cf. Hinde and Spencer-Booth, 1967, for the normal weaning process in rhesus macaques), though this problem has never been systematically investigated for langurs. Leaving (or being chased out of the troop) has never been observed before the juvenile stage was reached.

Prepubescent sexual solicitations of the male may be an option open to the older female infants. Scrapetail's 12-to-13-month-old daughter N-M presented to Toad from a distance of about ten meters and shook her head at 12:50 on January 8, 1974, just two hours after she was attacked and wounded by him. Similar solicitations were observed at 10:15, 3:00, and 5:14 on January 13 and at 8:00 a.m. on January 28.

Sexual solicitation (that is, presentation with head shudder) of Mug by 10-to-11-month-old Virginia was observed once, at 4:55 on September 10, 1972, the day after Mug had attacked and severely wounded another Hillside infant, Scratch. Whereas N-M often approached Toad to within three to five meters before presenting to him, Virginia solicited Mug only once, after he approached her.

Presenting is a normal posture for a subordinate to assume before a dominant animal, but frenzied head-shaking is not. Infant females were not observed to exhibit estrous signals in any other context. On rare occasions, subadult males were observed to present to other subadult males and mildly shake their heads (recorded on 16 mm film), but the frenzy characteristic of true estrous shuddering and of the estrous mimic of prepubescent females was absent.

Infanticide as a Primate Reproductive Strategy

Perhaps the most convincing argument in favor of the adaptiveness of infanticide is its prevalence throughout the natural world. In the years since I first went out to investigate the "bizarre" langur aberration, zookeepers and fieldworkers have compiled sufficient information to characterize the trait as widespread among primates. Adult male infanticide is known to occur among the suborder Prosimii and in each of the superfamilies of higher primates: the New World monkeys, Old World monkeys, and the Hominoidea—the great apes and ourselves. In the roughly 15,000 hours that behaviorists have watched hanuman langurs, we have documented the disappearance of some 39 infants coincident with male invasions. By comparison with *Presbytis entellus*, other colobines have scarcely been studied, yet even in preliminary work male take-overs accompanied by the disappearance of infants have been reported for three species and are suspected in two others.

The pattern of invasions by nomadic males, defeat and ouster of resident males, and the killing of their offspring is by no means unique to primates. Outside of the primates, lions provide the closest parallel. As with langurs, the stable reproductive unit is composed of matrilineal relatives encompassing several generations. According to George Schaller, these sisterhoods are closed social units, difficult for any alien female to join (1972a). Typically, male lions are forced out of their natal groups before reaching adulthood; if they ever rejoin the pride, it will be as nomads who actively expel resident males and appropriate their lionesses. At such times infanticide may occur (Schaller, 1972; Bertram, 1976). Occasionally male lions try, as Shifty Leftless did, to maintain jurisdiction over two prides at once.

Like lions and langurs, the terrestrial patas monkeys of Africa live in one-male harems surrounded by male bands (Hall, 1968). Other than keeping away other males, and occasionally diverting or chasing predators (Struhsaker and Gartlan, 1970), patas males do not participate in childrearing. Under conditions leading to high rates of male replacement (for example, high population density), one would expect infanticide to be no less advantageous to males of this species than it is among langurs or lions. Since infant-killing has not been previously reported among patas, this little-known species makes an informative test case for the hypothesis that infanticide is likely to evolve where there is intense competition between males for females and where the tenure of male possession of females is brief. Given this prediction, it is interesting that, like lions (Schaller, 1973) and langurs, patas monkeys exhibit estrous behavior when exposed to an alien male.

Although short tenure of access to females is apparently crucial to the evolution of infanticide, the benefits to males from infant-killing are not necessarily confined to breeding systems in which males usurp harems. As the work of Fossey among gorillas and of Goodall among chimpanzees continues, these Great Apes are checking in with an infanticidal propensity less marked than among the Colobinae but one which is nevertheless impressive when the comparatively small number of births among apes is taken into account. In some 7,000 hours of observation, Fossey witnessed three cases of adult male infanticide and inferred the occurrence of three others (1976). After Bygott's initial observation of infanticide among chimpanzees, at least three others have been reported by Goodall and others who continued to monitor events at Gombe (see also Suzuki, 1971). Yet neither gorillas nor chimpanzees have a social system resembling langurs.

Among the long-lived gorillas, dominant males remain in possession of their groups for many years. Young females move between these harems, successively picked up by various adult males. Once they reach the age of full maturity, however, gorilla females settle down and remain in one group (Fossey, 1976). Hence, while a dominant silverback male may have almost permanent possession of an older female, his average tenure of access to young, transient females will typically be short. It may be significant, then, that all known cases of infanticide among gorillas have involved primiparous young mothers. Similarly among chimpanzees, adult male infanticide has only been reported among females who are in transit —either mothers who have entered a strange locale occupied by unfamiliar males or females who have returned home pregnant after such a visit. In addition to the transitory status of these females, the long birth interval among apes intensifies the disadvantage for a male who allows a female only temporarily within his sphere to rear a competitor's offspring. Among all these species, captured or encountered females would have infants sired by some other male. By killing such infants, the male proportionally decreases the reproductive success of his competitors; by inseminating such females himself, he insures his own.

Infanticide among colobines dates far back in time. What are the implications of this infanticidal heritage for humans? There is little reliable evidence to support the hypothesis that human males have been selected to murder infants in order to increase their own reproductive success. Only in the realm of anecdote, literature, and the rumors of genocide that accompany war do we find alien males slaughtering infants who are not their own. Some of these examples are quite familiar to us, such as the wrath of Herod who "sent forth, and slew all the children that were in Bethlehem, and in all the coasts thereof, from two years old and under" (Matthew 2:16) or the command of Pharaoh that all Hebrew sons should die at birth

(Exodus 1:16). Among most preindustrial human societies, however, infant-killing is primarily practical. It is determined by economic constraints, the probability of infant survival, and future marriage potentialities (Dickeman, 1975; Langer, 1972; Laughlin, 1968; Konner, 1972; Mead, 1971; Wagley, 1969). Females either concur in this decision or are actually responsible for dispatching the ill-fated child. Occasionally a husband may kill an infant sired by some other male (among Eskimos, Balikci, 1970; among the Tikopia, Firth cited in Lorrimer, 1954; or among the Mundurucu, Murphy and Murphy, 1974). I encountered only one anecdotal description of the murder of children by invading males, reported for the Yanamamö tribe of Brazil (Biocca, 1971). More typically, Yanamamö mothers are kidnapped and their infants left behind, not killed. This, of course, is functionally equivalent to infanticide, since it results in the cessation of lactation and the probable starvation of the offspring. Yanamamö women themselves are well aware of these implications of abduction (Chagnon, 1968). Interestingly, Yanamamö society is characterized by raiding for women and by the tendency for a few males to be disproportionately represented in the gene pool of succeeding generations (Neel, 1970; Chagnon, 1972). As among lions and langurs, there is intense competition between males for females, a fast rate of political change such that the opportunity for males to reproduce is short—two to three years in the case of langurs, potentially ten but closer to two among lions, and until the next raid among the Yanamamö.

Scanning our entire species, nothing resembling a genetic imperative for infanticide can be found. Is this because we are different from other primates, more cultured, and more nearly the products of our upbringing, history, and beliefs than other primates are? Or simply because, in general, male tenure of access in humans is rather long by langur standards?

9 / Two Different Species

> *Laura:* "The mother was your friend, you see, but the
> woman was your enemy. Sexual love is conflict. And
> don't imagine I gave myself. I didn't give. I only took
> what I meant to take. Yet you did dominate me . . . I felt
> it and wanted you to feel it.
> *Captain:* You always dominated me. You could hypno-
> tize me when I was wide awake, so that I neither saw
> nor heard, but simply obeyed. You could give me a raw
> potato and make me think it was a peach . . . Just one
> thing more—a fact. Do you hate me?
> *Laura:* Sometimes—as a man.
> *Captain:* It's like race-hatred. If it's true we are des-
> cended from the ape, it must have been from two dif-
> ferent species. There's no likeness between us, is
> there?
>
> The Father by August Strindberg, 1887
> (from a translation by Elizabeth Sprigge)

The study of free-ranging monkeys is like a jig-saw puzzle poured
out from its box; the cover picture and a number of pieces have been
lost. The significance of isolated bits becomes apparent only after
much of the puzzle has been fitted together; some pieces remain
mysteries. Compromising between pure speculation and precise
records, the primatologist must summarize the lifeways of species.

In the life of any animal, physical conditions and capacities, opportunities, reproductive potential, and the number and condition
of close relatives vary greatly over time, and one would expect the
optimum strategy for fitness at each life stage to vary accordingly
(Fisher, 1958; Gadgil and Bossert, 1970). This chapter outlines the
typical life stages of male and female langurs, providing a general
framework for the specific strategies described throughout this
book. Because the information necessary to weight the costs and
benefits of different strategies is simply not available for langurs
—a species which has never been studied under anything approaching controlled conditions—many of the optimizing strategies
suggested here for each life stage are hypothetical. Additionally,
where specific details are not available, I have extrapolated from
better-known species to construct what is undoubtedly only a first
effort.

291

Growing Up a Langur

The fetus in utero absorbs nutrients from the mother via the placenta. Based on information from sheep, the chain of events leading to parturition at term apparently begins in the fetal hypothalamus (Liggins, 1972). At birth, the little control the infant had over its fate diminishes considerably.

The newborn infant is totally dependent on its mother for nourishment and on its mother or allomothers for mobility and security. In the first weeks, infants have two options: to cling for their lives to whichever female is nearest or to complain loudly at poor treatment and in doing so to attract another allomother or prompt retrieval by the mother. Despite the infant's preference for its mother, newborn langurs spend as much as 40 to 50 percent of their daylight hours with allomothers. Whereas an allomother highly motivated to retain the infant treats it with solicitude, less motivated babysitters soon tire of their charges and may abuse or abandon them. Under special circumstances, when she is faced with an infanticidal male who seems likely to eventually kill her infant, even a mother may desert her own infant. The dependent newborn has little recourse. With time, infants gain in independence and in control of their own fate (fig. 9.1). Despite their growing independence, infants may seek to prolong the rewards of dependence.

In a recent and important contribution to the study of relations between parents and offspring, Trivers (1974) points out the potential conflict of interest between an infant who holds its self-interest above all others and a parent who, all other factors being equal, values each offspring equally. Generally, the mother benefits most from terminating her infant's dependence on her, and, not surprisingly, it appears to be the mother who takes the most active role in weaning the infant (based on quantitative data for rhesus macaques provided by Hinde and Spencer-Booth, 1967, and on casual observation of langurs). According to the arguments advanced by Trivers, this sacrifice of nutriments detracts from the mother's overall fitness, hence it would be advantageous for the infant to prolong nursing much longer than it would be advantageous for the mother to provide milk. As an outgrowth of the mother-offspring conflict, fierce weaning disputes between mothers and infants are predicted, and, in fact, typical. Among langurs, maternal stratagems to discourage infants from suckling range from mild rebuffs (pushing the infant away, slapping or nipping it) to flight.

Infants respond to maternal rebuffs with noisy tantrums, pretended distress, or, occasionally, fierce retaliatory aggression (fig. 9.2). Mothers have the upper hand in weaning disputes; neverthe-

Fig. 9.1. Allomothers are an important part of the developing infant's social environment. In his earliest demonstrations of independence, Brujo (shown here at the age of three weeks) hopped on his own between his mother and other adult females. (The town of Abu can be seen in the background.)

Fig. 9.2. Weaning disputes occasionally entail fierce retaliation by the infant against the mother who has rebuffed it.

less, harassed mothers may give in temporarily before the fierce persistence of their offspring. Virtually no quantitative data are available on the onset and duration of weaning among langurs, or the environmental and social factors that affect its timing. Based on casual observation, langur mothers initiate weaning any time from six months onward, and this weaning may last for some time, usually not terminating until the infant is thirteen to twenty months old. Confronted with an infanticidal male, a mother may speed up the weaning process or abandon her infant altogether. If it forestalls the male from attacking the infant, early weaning would benefit the infant as well as the mother. Interestingly, as far as I can determine, infants resist such weaning with the same persistence that infants do under normal circumstances, suggesting that the maternal strategy has not quite filtered through to the offspring.

Observers of langurs seem to agree that young males spend more time chasing, wrestling, and engaging in dominance-oriented behaviors than young females do. By contrast, young females spend more time in association with adult females and infants. The direction of these differences is consistent with those found in a number of other primate species, including rhesus macaques (Hinde and Spencer-Booth, 1967; Harlow, 1971), vervets (Lancaster, 1971), and human children (Blurton Jones, cited in Harlow, 1971).

At Abu, the development of infants was not traced closely enough to say at what age differences between the two sexes first become apparent. Jay (1963) reports that at Orcha and Kaukori, she detected differences by the age of three to four months; males were rougher and more active in their play than females were. However, Sugiyama (1965a) notes that at Dharwar it was not until six months after birth that "slight differences" could be detected; males and females were still mixed in group play at the age of ten months. Such differences could be due to the monkeys themselves, to differences between observers, or, more probably, to differences in troop composition. That is, whereas Jay's main study group at Kaukori contained more than fifty langurs, the group studied by Sugiyama at Dharwar contained less than one half that many. With fewer alternatives to choose from, infants at Dharwar might have continued to play with members of the opposite sex longer than they would have when provided broader options.

Almost certainly, play provides vital practice for adulthood (Lancaster, 1971; Dolhinow and Bishop, 1972). One would scarcely expect an infant to forego play simply because the ideal partner did not happen to be available. Mixed play groups were common at

Abu. Nevertheless, when same-sex partners were available, it did seem that juvenile and subadult males preferred to wrestle with and chase other males. The "impulse to play," regardless of who the partners were, could be seen in the case of the lone Hillside infant, Miro. Because of the very high infant mortality in Hillside troop between 1971 and 1974, Bilgay's 10-month-old son was the only immature in the troop during the 1973-74 study period. When I first observed Miro, I was struck by how extremely timid and dependent on his mother he appeared to be. When the troop moved, even a short distance, Miro would hang behind and whine until Bilgay returned, took him up, and carried him.

Because I had developed an image of Miro as a timid "only child," I was surprised to observe that Miro's behavior with strange langurs was anything but timid. During encounters between the Hillside and Bazaar troops, Miro assertively sought out playmates of any age in the Bazaar troop. He would challenge parties of several alien adults and take on juvenile wrestling partners much larger and heavier than himself. Despite the trouncings Miro inevitably received, he remained undaunted. As far as I know, these contacts with Bazaar troop immatures were the only opportunities for play that Miro had.

For both sexes, the repertoire of langur play behaviors appears to have a high degree of relevance to adult activities and social relationships. On one hand, the fanciful and often exaggerated movements involved in play may be important in the generalized development of motor skills (color fig. 2). On the other hand, specific information gained from exploratory behavior and specific skills obtained in activities such as feeding, homosexual and heterosexual mounting, chasing, fleeing, leaping, grappling, harassing, and threatening other animals may be directly relevant to adult pursuits. Most importantly, perhaps, the maturing infant learns how to anticipate and evaluate the reactions of other langurs.

One of the most striking pastimes of immatures of both sexes, but particularly of infant and juvenile males, was derring-do—that is, an immature challenging a much larger animal by running up to him, lunging in place, grimacing, squealing, and in some cases contacting him. Young males frequently challenged members of other troops in this way during intertroop contacts. Within the troop, infant and juvenile males might challenge an alpha male by coming up behind him and pulling his tail or by directly charging him. Typically these tiny antagonists wear cheek-to-cheek fear grimaces, and are ignored by the challenged animal. Essentially the same behaviors are used in harassment of copulating couples. Given the importance of harassment, threat, and bluff in the life of both male and female langurs, such play behavior is in fact serious business.

In addition to learning motor skills and how to accurately assess complex environmental and social situations, any optimum strategy in prereproductive years must include avoidance of debilitating injuries. Play-fighting and real fighting are similar, but the elaborate facial expressions worn while playing, the exaggerated gamboling movements, and, of course, the small size (and possibly also color differences) of immatures may operate to alert other animals that these threats are not to be taken seriously. In this way, young animals are given license for behavior that otherwise might elicit damaging attack from a larger animal. In addition, conspicuous play faces and play posturing, such as inviting demonstrations of helplessness (lying on one's back, exposing a vulnerable cream-colored belly to be pounced upon), communicate to potential partners a willingness to play.

Infant, juvenile, and subadult females, like young males, play-fight and engage in play domination and threats. But an even more time-consuming pursuit for young females is taking and carrying infants. By the age of about ten months, females exhibit much greater interest in infants than males do; by thirteen months, or in some cases sooner, young females try to take infants whenever they are available. At Abu, immature females were responsible for the majority of attempts to take infants. The access of nulliparas to infants varied with age and, presumably, competence. Whereas older nulliparas succeeded in taking infants on 76 percent of the occasions that they tried, females younger than fifteen months succeeded only 34 percent of the time. Young females were more reluctant to attempt to take infants than the highly motivated older nulliparas, and this may be due to the discouraging effects of frequent rebuffs. This differential motivation may also reflect maturational factors. Either because they were able to, or because they chose to, older nulliparas kept infants for longer periods, ten minutes on average, roughly twice as long as the younger females.

Despite a poor success rate in obtaining infants, younger nulliparous caretakers do improve with age. Improvement could be due to learning, maturation, or both. In any event, prior to their first pregnancy, most nulliparas have become skillful at carrying infants and keeping them quiet and contented.

The optimum strategy for nulliparas is to take and hold infants for relatively long periods and to treat them with solicitude, both in order to learn maternal skills and to assure that the complaints of the "practice" infants will not attract other allomothers or cause the real mothers to retrieve their offspring. These immature caretakers learn the appropriate responses to various infant-threatening contingencies and acquire other skills vital to the well-being of helpless young. Because newborn infants are only sometimes present in the troop, nulliparas only rarely achieve a level of expertise where further practice caring for infants would be redundant. If there are many nulliparas and few infants, access to charges may be competitive. If there are no newborn infants currently available within the troop, nulliparas may aggressively set about stealing an infant from a neighboring troop.

By the time a female delivers, her physiological investment in that infant is enormous. The female is under pressure then to practice during her prereproductive years so as to increase the survivorship of the infants that will be born to her. In addition to gaining experience in caretaking, young females benefit (more so than most young males) from gaining specific information about possible dangers and the availability of resources within the troop's home range where the female will permanently reside and rear young. The different perceptions of their environments by males and females provide a fruitful area for future research.

Beginning in preadulthood, females compete with troopmates for perquisites of survival and reproduction. Young females become increasingly active in displacement behaviors, and as a result of their assertiveness, these impetuous young females are more frequently displaced by higher-ranking animals than are older females who avoid conflicts. Eventually these young females rise above their elders in the displacement hierarchy. By the time their first or second infant is born, most young females have achieved a position in the top half of the female hierarchy. It may be advantageous for these young females to behave selfishly during their peak reproductive years, focusing almost exclusively on individual fitness. Because of the high degree of relatedness between female troop members, old females of much lower reproductive value would also stand to gain in inclusive fitness by acquiescing.

Young males who remain in the troop follow a similar schedule, beginning just prior to adulthood to improve their position in the troop's hierarchy. By the time they are four to five years of age, males are able to displace all adult females in the troop. Such males rank just below the alpha male and any other full adult males who might be in the troop. Only one troop studied at Abu (the Bazaar troop in 1971) had more than one resident adult male, and in this case, the oldest male clearly had prior access to positions, scarce resources, and estrous females. Generally speaking, however, overt antagonism between these males appeared mild.

By the time a young male is about four years old, canine teeth have erupted and the testicles have descended. For some years, however, both teeth and testicles remain smaller in size than those of a fully adult male. Adolescent males are apparently capable of copulating and have been observed to enter neighboring troops to steal copulations. There is a very low likelihood, however, that a young male who has not yet achieved his full weight and fighting potential could maintain exclusive sexual access to a troop of females, nor could he successfully defend against marauding males the infants that he might father.

Given these circumstances, the low level of antagonism shown by young males towards older males (in many cases their fathers) is adaptive. Measured in terms of food resources, security, and clandestine access to estrous females, it is more advantageous for a subadult male to submissively remain in a bisexual troop than to risk being driven from the troop by an older and larger competitor. Furthermore, it seems possible (but it has never been documented) that a maturing male who remains in his bisexual troop may one day inherit it. If so, it would behoove a young male to assist an alpha male who tolerated him in defending the troop from extratroop males who almost surely would not; the young male as well as the alpha male would be evicted in the case of a take-over. Varying relationships between young males and the troop's alpha male might explain the difference in participation by young males in intertroop encounters and in defense of the troop against invading males.

Young males who are evicted from their natal troops, either by their own fathers or by usurpers, join with other males in nomadic all-male bands. In some cases, males ousted en masse remain in association with their father and other male relatives (fig. 9.3). The degree of relatedness between these males might influence the extent to which they would cooperate in invading bisexual troops. Nevertheless, until they are fully grown, young males would have little chance of profiting from invading or usurping troops, and injury during an invasion might well jeopardize future opportunities. Young males have little incentive, therefore, to take the same risks in fighting that older animals do.

Because a young male may either be allowed to remain in his natal troop or be forced to join an all-male band, the age and sex of his associates during adolescence might vary considerably. In either event, the subsequent reproductive opportunities of this male

Fig. 9.3. Splitear, his eight sons, and two females with offspring temporarily traveling with the ousted males continued to police their former home range against encroaching male bands. Here, Splitear (large male in foreground) grinds his teeth just prior to an encounter with the Waterhouse band.

will depend to a large extent on his ability to compete with, to dominate, to bluff, and to deceive other males. Not surprisingly, the social orientation of all males, be they in troops or bands, is toward other males.

When a subadult male steals a copulation, he may have more at stake than a possible insemination. The occasional failure of subadult males to mount females properly suggests that practice may be an important component for successful mating (fig. 9.4). Mounting of nonovulating females may have some advantage, then, for inexperienced males. If males benefit from sexual behavior even when there is no possibility of insemination, then, interestingly, male sexuality is not necessarily dependent on the proximity of an estrous female. At Abu, only young adult males were ever observed to masturbate; any ejaculate produced was eaten.

Learning may also be relevant to female sexual behavior. Consort relations between the third year when a female begins to cycle and the fourth year when she gives birth may provide useful experience.

Fig. 9.4. Subadult male invaders from Splitear's band entered School troop to steal copulations from an estrous female there. They are harassed by juveniles and adult females from School troop. Note that the male is incorrectly mounted. A second invader waits just behind him, while the third lunges at a harasser (below the mounted pair).

In 1972, for example, a subadult female, Scrapetail, in the Toad Rock troop came into estrus. Whenever the alpha male, Splitear, approached her, she ran away and again frantically solicited him from a distance. This alternating courtship and flight lasted more than three hours without a single successful copulation. Soon after, however, Scrapetail must have conceived; in January of 1973 she gave birth to her first infant.

No overt sexual behavior was ever observed among very young females, with the exception of pseudoestrous solicitations by endangered infants. Prepubescent estrous solicitations were observed in the case of only two individuals: 12-to-13-month-old N-M and 10-to-11-month-old Virginia.

The Reproductive Years

For four years in the case of females, and closer to six for males, langurs develop physically and acquire skills necessary for survival and reproduction. Among males, opportunities to reproduce will largely depend on competition with other males. Among females, competition with other females is only a part of the large investment necessary for the rearing of offspring.

"The stag company is a number of egocentric males and is a very loose organization" said Fraser Darling (1937) in his classic study of red deer; no less so are bands of langurs. Options open to males in bands by and large include haunting or skulking about a troop, attacking it and then retreating, temporarily joining it, or taking it over. Benefits obtainable and the costs that might be incurred differ with each strategy and with the circumstances in each case; the balance of costs and benefits will be especially affected by the reproductive value (usually a function of age and condition) of the male involved and of his opponents. For example, the risks taken by Shifty Leftless, an extraordinarily audacious older male, may have been devalued by his age and correspondingly lower reproductive value. No comparable devaluation reduced the risks of the young adult males who fled before him.

Dominance relations between males are not always so clear-cut. If the number of extratroop males engaged in an attack on a bisexual troop increases their likelihood of success, it might be disadvantageous for a very powerful male to display his superiority prior to the attack. By so doing, he might discourage other members of the band from participating. For this reason, the dominance hierarchy visible in a male band before a take-over might not predict which male will remain in sole possession of the bisexual troop afterwards. Because members of the male band will not share equally the benefits of a troop take-over, extratroop males should show variable commitment to attacks on troops. This variation should be apparent both in the intensity with which certain males fight and in the fluctuating composition of male bands.

Once a male from outside the troop has taken over a troop, the gain to him from killing unweaned infants in the troop can be calculated in terms of reproductive time gained. The interval between births represents the sum of: 13-20 months (birth to weaning) + whatever lag + 6.5 months gestation. The benefit gained by a usurper from killing an infant will be in inverse proportion to how far along the infant is in the weaning process. That is, the time gained will be roughly 13 months minus the age of the infant. Hence, the attack by Toad on N-M, an infant female already 13 months of age in January of 1974, was potentially less beneficial to him than attacks on younger infants would have been. Furthermore, should Toad remain in the Toad Rock troop for an additional three years, N-M would then increase the ranks of reproductively active females available to him.

The exigencies of short tenure of access to females favor the evolution of infanticide, but it is important to note that prior adaptations are essential before males can benefit significantly from killing infants. In particular, the flexible breeding physiology that permits females to conceive again shortly after the death of a suckling infant is crucial for the secondary male adaptation of infanticide. In the case of langurs living in areas with long male tenure, or, to take a hypothetical example, in areas so harshly seasonal that infants could only be born during a circumscribed birth season (such as spring), one would expect the conditions favoring infanticide to be diminished, if not altogether absent. If traits carried by infanticidal males were for some reason disadvantageous to progeny, females who *failed* to boycott inseminations by infanticides would themselves be rapidly selected against. As with other male behaviors such as ranging patterns, infanticide is overlaid upon prior female adaptations. Infanticide is not in itself advantageous. In order for the trait to spread, conditions favoring it (and these must involve female physiology and choice) must prevail long enough for the progeny of infanticides to prosper.

Compared to other primates such as baboons, marmosets, Japanese macaques, or humans, in which males may play an important role in childrearing and defense, langur females are remarkably self-sufficient. Apart from insemination, the only essential function fulfilled by adult males is to prevent other males from invading the troop and killing infants. At Abu, females in Hillside troop spent roughly 25 percent of daylight hours in 1972 and 1973 with no adult male present. During February and March 1973, adult males almost never spent the night at the same site with these females. The Hillside females suffered no immediate ill-effects other than vulnerability to alien males.[1] The troop foraged as usual and even engaged in intertroop conflicts with the Bazaar troop, once while the Bazaar alpha male, Shifty, was also off, chasing males from Hillside troop.

The responsibility of childrearing falls on females. By the time her first infant is born, a young female has achieved an advantageous position within the female displacement hierarchy and has had extensive experience handling newborn infants. Because of the dire reproductive consequences for a mother unresponsive to the needs of her dependent offspring, one would expect females generally, and in particular pregnant or lactating females who are most "at risk" in this respect, to respond positively to newborns. When a new infant is born, most or all females in the troop inspect it. Nevertheless, only some females, that is nulliparous, pregnant, or lactating females, repeatedly take and carry the infant. While pregnant or lactating females might benefit from holding infants, the benefits soon become redundant; females who have had prior experience rearing infants tend to keep infants less than half as long as nulliparous caretakers. Presumably multiparous mothers have less to gain from expending energy on an infant that belongs to another female than they would by caring for their own, such females soon lose their motivation to retain the infant and may discard it or abuse it in an effort to get rid of their clinging charge.

Unquestionably a langur mother is able to recognize her own infant. A mother holding two infants simultaneously may desert the second infant, but never her own. Lactating allomothers do not normally suckle offspring other than their own. To do so would divert resources from their own current or future offspring. Nevertheless, in at least one instance a mother whose infant had died was known to adopt the offspring of another female (page 88).

Outside of the mother-infant relationship there is little evidence that females discriminate between infants on the basis of relatedness. Nevertheless, allomaternal caretaking may be a cooperative venture insofar as the females most motivated to caretake are those

1. It is not known, of course, how male absence over a long period of time would affect a troop and its range. Events at Jodhpur suggest that maleless troops might be at some disadvantage in defending territory (Mohnot, 1971b). Nevertheless, it is worth noting that despite multiple changes in male "leadership," langur home ranges and patterns of habitat use remain fairly constant; it must be assumed that the responsibility for such traditions resides with females.

with something to gain from caretaking and insofar as caretaking means foraging freedom for the mother. Except in emergencies, an infant does not appear to benefit from the attentions of allomothers and indeed may suffer from them. In this respect it may be significant that on average caretakers will be more closely related to the mother than they are to her offspring. Because of the langur pattern of male take-overs, troop members are likely to be more closely related to individuals born about the same time that they were, who will share the same father.

Females of high reproductive potential may benefit from selfishly monopolizing prerequisites while leaving matters of troop defense in the hands of their older female relatives. Though all females, including subadults and young adults, participate in troop defense, older females often take the most active role. Older females also initiate troop movements; in doing so, they presumably share their greater knowledge of the home range and of transient resources within it with other troop members. At Abu, alpha males also occasionally determined troop movement. At Dharwar and Kaukori (Sugiyama, 1965a; Jay, 1963c), however, leadership by males was the rule. This discrepancy in the identity of troop leaders may reflect differences in the social conditions in the troop observed. Because of my observational bias towards reproductively significant events, I tended to focus on troops in which new males had recently usurped control. Hence during many of my observation hours I was watching alpha males who were relative newcomers in their troops. These new males were almost certainly not as familiar with the troop's home range as were the permanent female members. By contrast, in the case of both the Dharwar troop (the 30th troop) and the Kaukori troop, a look at troop compositions suggest that these two troops had been stable for some years. Hence the males may have been almost as familiar with the terrain as many females were. From these examples, it appears that old females may play a greater role in leading the troop during transitional periods between the tenure of one alpha male and the next.

Langur life histories presented here illustrate the extent to which the two sexes differ. The demands of reproduction have led to the evolution of two quite different creatures, two sexes caught in the bounds of irreconcilable conflicts. In only a few areas will the self-interests of consorts overlap.

Concurrent and Conflicting Interests between the Sexes

Whereas female troop members are closely related to one another, they will be only distantly related to their consorts, the alien males who usurp their troop. Within six to seven months of a take-over, however, both the new male and the females within the troop should have relatives in common among the offspring born in the troop. The individual and inclusive fitness of both the male and his harem are linked to the survival of the troop's immatures. The well-being of these offspring inspires a certain consensus regarding the defense of the troop and the troop's resources. Furthermore, both alpha male and troop females benefit from a system in which fe-males with the highest reproductive value have preferential access (among females) to resources and take the smallest risks in troop defense.

Since invading males might kill infants and drive out their adoles-cent male relatives, females in the troop would be expected to co-operate with the current alpha male in prolonging his tenure.[2] However, if the troop had been stable for a period of five or more years, adult females might be closely related to adult males in the troop and might encourage inseminations from outside males rather than breed with their fathers.

The survival of offspring born in the troop, defense of the troop's resources, and fostering the maximum reproductive potential among the troop's females are the main areas in which the selfish interests of adult females and the alpha male might overlap. Never-theless, even within these three areas there is room for disagree-ment. The shorter the tenure of the male, the less would his inter-ests and those of troop females coincide. There are four main areas of conflict between males and the females they seek to possess.

Troop membership. Because of the varying degrees of relatedness between troop females and outsiders and between the incoming alpha male and immatures already present in the troop, one would expect troop females and the alpha male to disagree on the optimum composition of the troop. Since immature males sired by a previous alpha male would compete with a new leader's own offspring and harem, the new male might benefit from evicting them. Whether or not females would benefit from the ouster of their sons and nephews is unclear. Attempts by the alpha male to merge one harem with an-other highlight a more clear-cut conflict of interest. It would be to the male's advantage to gather under his protective custody as

2. Actually, the consensus among females in the troop concerning take-overs may vary more than this statement implies. For example, a subadult about to enter her reproductive career might "prefer" a take-over by a young adult male rather than a continued but vulnerable reign by an old or disabled male unlikely to last much longer. Benefits to the young female would have to be weighed against the possible costs of a take-over for a troop full of relatives.

many reproductively active females as possible. In three of four cases observed at Dharwar and Abu where males attempted to combine two troops under a single leadership, the attempts were foiled by antagonism between females in the two troops. In the single successful case, the merger was apparently facilitated by the eagerness of females in one troop without infants to gain access to infants in another troop (p. 153).

Female opposition to mergers is at the root of a fundamental dilemma for the langur male who is able to take over more than one troop; if he divides his time between more than one troop, he may sire more offspring than he can protect from marauding males. Essentially, the male must choose between remaining in a usurped troop and moving on. If he does remain, should he concentrate his energies there, or attempt to expand his sphere of control over more than one troop? In the case of three usurpers in Hillside troop between 1971 and 1975, the males unanimously opted for the double-usurper strategy. In contrast, Harelip, the sole usurper of School troop during this period, stayed with a more conservative one-troop-only policy.

Allocation of troop resources. One explanation for the opposition of females to troop mergers could be the matrilineage's stake in preserving for its own use home resources. Females in an area of seasonally scant resources might jeopardize their own fertility or the survival of their offspring by overexploiting this legacy. The possibility that two joined troops would then have a joint home range, the size of both home ranges put together, has never been investigated, but there may well be a limit to the size of the territory that a troop of langurs can successfully defend. Quantitative information on the effects of harem size on the fitness of individual members is not available for any of the nonhuman primates. Nevertheless, information for several species of birds and mammals suggests that female fitness is affected by harem size (among marmots, Downhower and Armitage, 1971; various birds, Orians, 1969; and humans, Gomila, 1975). The direction of the effect presumably depends on the availability of food and other vital resources.

A similar argument might be advanced to explain the apparent conflict of interest evidenced in sexual harassment of copulating couples by adult females. It is possible that females might be attempting to delay conception among female troop members whose offspring would then compete for resources with their own.

A third and more obvious disagreement over the allocation of resources within the troop involves preferred food items and feeding positions. Invariably the alpha male obtains the lion's share and lion's usage of resources through his ability to displace all other troop members. Since the male's fitness is linked to that of the females in his harem, the only explanation as to how his despotism might be adaptive would be that his fitness is more affected by his own physical condition (and hence his ability to control his harem) than it is by any decrease in the fertility of his females due to reduced access to preferred foods.

Adultery. If it becomes likely that an alpha male will lose control of his troop to extratroop males, he and his harem will not "agree" on how cycling females should behave. In such cases an ovulating female should seek to establish consort relations with as many males as possible, and in particular those males likely to take over the troop. The alpha male, on the other hand, should seek to retain exclusive sexual access. If, however, the female is already pregnant by the alpha male, then both parties would benefit from a policy of soliciting other males. This assessment, if correct, could explain Mug's tolerance of copulations between Hillside females suspected of being pregnant and the five young invaders temporarily traveling with Hillside troop.

One of the more interesting aspects of copulations by females with outsiders is the possible role of adultery in incest avoidance. The single troop at Abu that remained stable over a period of five years was also the troop with the highest incidence of adultery. Whereas the young females in this troop avoided incestuous copulations with the alpha male (their putative father), there was no indication that he did. Presumably, a female would have more to lose, proportional to her lifetime's reproductive success, from bearing a defective offspring than the male would.

Timing of conception. Because an alpha male's tenure in a troop is on average short, his fitness is not linked to the overall reproductive success of the females in his harem during their lifetimes but only to their fertility during his tenure. If he could, it would be to the male's advantage to telescope a female's entire reproductive career into his brief span as alpha male. The optimum production of healthy progeny by the female, however, is not compatible with such a compressed schedule.

Prior to the birth of an infant, a female (or rather her physiology) appears to have control over her schedule of conceptions. After birth, however, males may drastically distort options open to a female by killing her infant. If physically possible, the most advantageous course for a female at this juncture would probably be to conceive again. It is not known whether a female could take into account an infant's chance of surviving prior to conceiving it; the difficulty among langurs of predicting social conditions seven months in advance would seem insurmountable.[3] At the very least, however, a female might avoid insemination by a very old or by a disabled male—a class of individuals that was never observed in the bisexual troops at Abu.

Rarely do the best interests of the female langur coincide with those of her consort. As Strindberg so clearly perceived, sexuality means conflict. Apart from insemination, langur females have little use for males except to protect them from other males. The outstanding question, then, is why females should tolerate males at all. Why allow themselves to be subjected to the tyranny of warring polygynists? Evolution operates on so vast a time scale that alternatives have been open to female langurs for many thousands of years. Large body size, muscle mass, and saber-sharp canines might just as well have been selected among females as among males. Why should not a female weigh the 18 kilograms that a male ordinarily does? Alternatively, female relatives could ally themselves to a much greater extent than they do. The combined biomass of three females operating as a cooperative front against an infanticidal male surely should prevail. Since infanticide depends for its evolutionary feasibility on the insemination of the female after the death of her infants, females could fail to ovulate in the company of an infanticide.

The fact that females do not grow so large as males, that they do not selflessly ally themselves to one another, and that they do not boycott infanticides suggests that counter-selection is at work. Once again the pitfall is intrasexual competition—this time competition

3. Nevertheless, among several species of small mammals, female physiology does appear to be sensitive to some sort of "calculation" concerning the social conditions likely to be operative at the time her next infant is due to be born. In the case of mice and voles, for example, pregnant females spontaneously terminate pregnancy when confronted with an alien male (Bruce, 1960; Stehn and Richmond, 1975). It is possible, but by no means conclusively demonstrated, that at least one species of primate—*Homo sapiens*—responds to a fetus unlikely to survive by either aborting it (Roberts and Lowe, 1975) or by reducing prenatal investment (Baird, 1945; Drillien and Richmond, 1956). The additional possibility that langur females are adjusting the sex ratio (Trivers and Willard, 1973) according to social conditions at conception must await a larger sample of infants than is presently available. It is possible that female offspring would be a better risk than males at times of political instability. It is suggestive, but nothing more, that of 53 births following male takeovers at Dharwar and Abu, only 40 percent of the infants born were male.

among females themselves for representation in the next generation's gene pool. Whereas direct competition between males for access to females selects for males who are as big and as strong or stronger than their opponents, a female who "opted" for larger body size in order to fight off a male might not be so well adapted for her dual role of both survivor (of drought and other climatic fluctuations) and childbearer. An oversized female might produce fewer offspring than her smaller cousin. In the long-term evolutionary scene, her cousin's progeny would prevail.

Intrasexual competition is lessened by the close genetic relatedness among female troop members, but it is by no means eliminated. A female in her reproductive prime who altruistically defended her kin despite the cost to herself might be less fit than her cousin who sat on the sidelines. Similarly, it would be to her progeny's detriment that a female refused to breed with an infanticide; if possession of the trait is indeed advantageous, her sons would suffer in competition with the offspring of less discriminating mothers. Apart from stop-gap strategies, the only viable option open to females is the energetically less costly, and less risky, strategy of deceit, to convince a usurper that he is the father of her subsequent offspring, to make him take a raw potato for a peach.

For generations, langur females have possessed the means to control their own destinies; caught in an evolutionary trap, they have never been able to use them.

Appendix 1 / Plants Used by Langurs

Based on samples collected at the top of Mount Abu and at Chippaberi and identified by Dr. K. M. Dakshini and by Dr. Steve Berwick. S = used as sleeping site; F = primarily fruit or berries taken; s = seeds or nuts taken. Note: Unless otherwise specified, flowers, buds, or leaves were taken as food.

Ficus bengalensis (S, F)
F. rumphii (F)
F. arnottiana (F)
F. religiosa (S, F)
F. infectionia (?)
Albizzia lebbeck (pods, leaves)
A. procera
Mangifera indica (S, F, flush)
Erythropia colorata (leaves, sap)
Eugenia jambolana (F, s)
Zizyphus jujuba (F, leaves)
Bauhinia racemosa (F, flowers, leaves)
Boswellia serrata
Morus alba
Morus (i. d.?)
Ehretia aspera
Flacourtia sepiaria
F. cataphracta
Bambusa frondosa
Bambusa sp.
Impatiens balsamina
Pogostemon plectranthoides
Poinsettia*
Cirewia sp.
Hibiscus sp.
Lantana camara (F)
Anogueissus sp.

Carissa spinarum [F, flowers, leaves]
Ingadulcis sp.
Cyanotis axillans
Crataera nurrula (i. d.?)
Grevillea robusta (S)
Prunus amyygdalis* (F, leaves)
Jasmium humile*
Terminalia bellerica
Citrus lemon*
Rosa muschata*
Jacaranda mimusifolia (pods, leaves)
Aegle marmelos (F)
Sesbania grandiflora
Eriobotrya japonica
Veronia cinerca
Ehretia laeuis
Tecoma stans
Cyanotis axillaris
Acer negundo
Ipomoea sindica
Ricinus communis
Gmelina (i. d.?)
Euphorbia sp.
Semecarpus (i. d.?)
Desmodium
Melothria maderaspatana (F)
Erythrina blakei

*Found only in the gardens of Abu town.

Appendix 2 / Weights of Hanuman Langurs

at Mount Abu, Jodhpur,

and Melemchi

TABLE A. Weights taken from Bazaar, Toad Rock, and School troops at Mount Abu, Rajasthan, during May and June of 1975. Except where otherwise indicated (in the righthand column below) all animals were weighed once, in the morning between 7:00 and 9:30 a.m. A Hanson heavy duty spring balance, 100 pounds by one pound, was used as illustrated in figure 6.10. As a rough approximation of how nearly full-grown they are, females are divided into three categories: multiparous, primiparous, and nulliparous. Females who are not yet cycling are classified as juveniles.

Category	Troop	Identification	Mean weight		Number of times weighed where more than once, and weights taken after 9:30 a.m.
			lbs	kg	
Adult male (mean weight, 18.4 kg)	Toad Rock	Toad	39	17.7	
	Bazaar	Righty Ear	41.3	18.7	2; once at 7:30 (41), and again at 3:50 p.m. (41.5)
	School	Harelip	41	18.6	
Multiparous female (mean weight, 11.4 kg)	Bazaar	Elfin	27	12.3	
		Overcast	23	10.5	
		Kasturbia	24	10.9	3:50 p.m.
		Quebrado	24	10.9	
		Short	24.5	11.1	2; once at 9:30 and 11:45
		(Wolf + newborn)	25[a]	11.4	

Category	Troop	Identification	Mean weight		Number of times weighed where more than once, and weights taken after 9:30 a.m.
			lbs	kg	
	Toad Rock	I.E.	25.5	11.6	2
		Pawlet	25	11.3	2
		Mole	26	11.8	6:05 p.m.
		P-M.	26	11.8	
	School	(Tamora + newborn)	26[a]	11.8	
		Adult #1	24	10.9	
		Adult #2	26	11.8	
Primiparous female (mean weight, 9.0 kg)	Bazaar	Junebug	18	8.2	3:50 p.m.
		Guaca	18.5	8.4	
		Breva	23	10.5	
Nulliparous female (mean weight, 7.9 kg)	Toad Rock	Hauncha (ca. 49 months)	21	9.5	10
		Pandy (ca. 52 months)	20	9.1	4
		No. 4 (ca. 52 months)	17	7.7	5
	Bazaar	Nullipara #1[b] (ca. 30 months?)	14	6.4	
		Nullipara #2[b] (ca. 30 months?)	15	6.8	
Juveniles (mean weight, 6.1 kg)	Toad Rock	N-M (ca. 29 months)	13	5.9	7

Category	Troop	Identifica-cation	Mean weight		Number of times weighed where more than once, and weights taken after 9:30 a.m.
			lbs	kg	
		Niza (ca. 30 months)	15	6.8	5
		Vert (ca. 19 months)	12	5.4	7
	School	Juvenile male	15	6.8	
		Juvenile male	12	5.4	
Infants (mean weight, 2.9 kg)	Bazaar	Guat (16 months)	6	2.7	
		Brief (13 months)	7	3.2	

a. Females carrying newborns are omitted from averages.

b. Though still quite small, both Nullipara #1 and Nullipara #2 exhibited estrous behavior.

TABLE B. Weights taken from troops B-6 and B-7 at Mandore Gardens, Jodhpur, Rajasthan, in cooperation with Dr. S. M. Mohnot. Weights were taken between 8:30 and 9:30 and 7:00 and 8:30 on June 21 and June 22 respectively. Mandore Gardens is a well-watered artificial park where langurs are provisioned by humans several times a day. Weights for this population probably approach the maximum for semi-desert-dwelling langurs during this particularly dry (1975 drought) month of June.

| Category | Location | Mean weight | | Additional notes |
		lbs	kg	
Parous female	Mandore	22	10.0	
(mean weight,	Gardens	22.5	10.2	
11.7 kg)		26	11.8	8 years old[a]
		24	10.9	
		29	13.2	
		27	12.2	
		30	13.6	Very old; very
		25	11.3	pregnant[a]
Nulliparous	Mandore	18	8.2	
female	Garden	22	10.0	
(mean weight,		15	6.8	
8.6 kg)		21	9.5	
Juvenile male	Mandore	11.5	5.2	Ca. 15
(mean weight,	Garden			months[a]
5.0 kg)		8	3.6	Ca. 13
				months[a]
		13	5.9	Ca. 15
				months[a]

a. Personal communication from Dr. S. M. Mohnot.

TABLE C. Weights taken from dead specimens collected in the Village Forest near Melemchi, Nepal in 1967-68 by Dr. Richard Mitchell who was studying the ectoparasites of Nepalese mammals (cited in Bishop 1975b).

Category	Mean Weight		Additional Notes
	lbs	kg	
Adult male	43	19.5	Collected November 3, 1967
(one weight,	32[a]	14.5	
19.5 kg)			Collected November 3, 1967
Adult female	34	15.4	Collected August 1, 1968
(mean weight,	37	16.8	
16.1 kg)			

[a]Apparently not a full adult; omitted from mean.

TABLE D. Comparison of average weights from Abu, Jodhpur, and Melemchi.

Category	Abu		Jodhpur		Melemchi	
	lbs	kg	lbs	kg	lbs	kg
Adult male	40.4	18.3	—		43	19.5
	(N=3)				(N=1)	
Adult female	25	11.3	25.7	11.7	35.5	16.1
	(N=11)[a]		(N=8)[b]		(N=2)	

a. Multiparous females only.
b. Parous females only.

Appendix 3 / A Numerical Method for Estimating Coefficients of Relationship in a Langur Troop, by Jon Seger

The average coefficient of relationship (r_{ij}) among the members of a social group is expected to influence the evolution of any trait that affects the average fitness of group members other than the bearer of the trait. It is usually impossible to write a general expression for r_{ij}, because it depends on a large number of variables, including the means and variances of group size, age-specific reproduction and mortality for males and females (l_x and m_x schedules), and migration rates (both into and out of the group, for males and females, by age), as well as other features of the mating system such as patterns of assortative or disassortative mating. Fortunately, tractable numerical methods are easy to devise for some special cases. This appendix describes a method that was used to estimate r_{ij} among the adult females of a hypothetical troop modeled after the langurs of the Mount Abu region.

Simplifying Assumptions   Time is reckoned in two-year "epochs." The troop always contains eight adult females, each two years older than the next younger; at the beginning of an epoch they are 5, 7, 9, 11, 13, 15, 17, and 19 years of age. Two years later, at the beginning of the next epoch, the eldest female (now 21) dies, and a female aged 5 years enters the adult population out of the pool of juveniles belonging to this troop. Thus the mother of the new recruit must now be aged 11, 13, 15, 17, 19, or is deceased (21, 23, or 25). This corresponds to an extreme form of "type 1" survivorship, in which all female mortality takes place before and after the ages of reproduction. Although not strictly realistic, such an age distribution appears to fit adult female langurs surprisingly well, and allows the calculation of all intratroop coefficients of relationship (r_{ij}) by the very simple procedure described below. The probabilities that each eligible female is the mother of the recruit were arbitrarily set as follows:

$$p(11) = 0.20$$
$$p(13) = .20$$
$$p(15) = .20$$
$$p(17) = .15$$
$$p(19) = .10$$

$$p(21) = .10$$
$$p(23) = .05$$
$$p(25) = 0 \tag{1}$$

These are proportional to age-specific fertilities five years previously if infant and juvenile mortalities are not heavily dependent on the mother's age. Because $p(25) = 0$, the eldest female alive in the troop is nonreproductive.

Juvenile males are assumed to leave the troop before reaching adulthood and are ignored in the model. Incoming adult males may remain in the troop 2, 4, or 6 years, and therefore father one, two, or three surviving female offspring. Different distributions of male tenure were employed, and these significantly affected r_{ij} among the females, as was expected. A critical assumption is that *adult males are unrelated to females and to each other.* This may be unrealistic, since the actual amount of inbreeding at Mount Abu, if any, is not known.

The Model   Construct a 10-by-10 matrix **R** with entries r_{ij}. Let the subscripts index females by their ages (5 to 23 in steps of 2) and the entries be their coefficients of relationship. Then $r_{ii} = 1$ and $r_{ij} = r_{ji}$; the matrix is symmetric. The population is progressed at each two-year "epoch" as follows. First advance each female one age-grade and introduce the new female in the upper lefthand cell.

$$r'_{i+2,\,j+2} = r$$
$$r'_{5,5} = 1 \tag{2}$$

where primes indicate elements of the matrix being formed out of the previous (unprimed) matrix. Then select one from among the females eligible to be the mother of the incoming female, according to the distribution of probabilities (equations 1). Call this female i. It follows from the rules for calculating coefficients of relationship in diploid species that the new female is related through her mother to every other female by

$$r'_{5,j} = \tfrac{1}{2} r'_{i,j}$$
$$r'_{j,5} = r'_{5,j} \qquad (j = 7, 9, \ldots, 23) \tag{3}$$

which completes the upper row and the lefthand column. If the new female's father was also the father of the previous recruit,

$$r'_{5,7} = r'_{5,7} + \tfrac{1}{4}$$
$$r'_{7,5} = r'_{5,7} \tag{4}$$

where "$=$" is understood to mean "replace the lefthand side with

the righthand side." If he was the father of the *two* previous re-
cruits, it is true in addition that

$$r'_{5,9} = r'_{5,9} + \tfrac{1}{4}$$
$$r'_{9,5} = r'_{5,9} \quad . \tag{5}$$

The paternal (half) sibling relationship between females 7 and 9
was established at the previous epoch by equations 4, and since by
assumption successive males are unrelated, the matrix is complete.

At each epoch, $\bar{r}_{ij}$ among living adult females is simply the average of the 56 off-diagonal elements of the 8-by-8 matrix derived from
R by deleting the last two columns and rows (21 and 23). For the
sake of simplicity, and to distinguish it from other averages that will
be mentioned, the $\bar{r}_{ij}$ particular to a given epoch will subsequently
be referred to as $\bar{r}$. Thus

$$\bar{r} = \frac{1}{N(N-1)} \sum_{\substack{j \neq i \\ i \neq j}} r_{ij} \tag{6}$$

where N is the number of females in the troop (here $N=8$).

The matrix contains additional information of potential interest.
In particular, $\bar{r}(i)$ can be calculated, giving a female's average relatedness to her troopmates as a function of her age.

$$\bar{r}(i) = \frac{1}{N-1} \sum_{j \neq i} r_{ij} \tag{7}$$

Also, the average relatedness of two females as a function of their
difference in age (D) can be found.

$$\bar{r}(D) = \frac{1}{N-D/d} \sum_{i=a}^{b-D} r_{i,i+D} \tag{8}$$

where d is the difference in age of successive females (here 2
years), a is the age of the youngest female (here 5 years), and b is
the age of the oldest living female ($b = a + d$ $(N-1)$, here 19 years).

Results   If this method works correctly, the *cumulative* average
r_{ij} should converge to definite limits as the number of successive
epochs through which the population is progressed approaches in-
finity. In particular, the cumulative r (that is, $\bar{r}$), should approach a
fixed limit equal to the expected coefficient of relationship $E(r_{ij})$
between two females who know nothing about each other except
that they belong to the same troop. As figure A shows, this convergence rapidly occurs from either of two extreme initial distributions

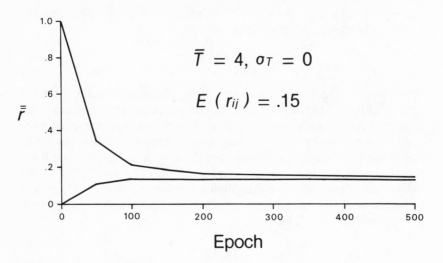

Figure A. Convergence of $\bar{\bar{r}}$ to $E(r_{ij})$. This graph shows the cumulative average $\bar{r}_{ij}$ within the model troop as a function of time, for one of the four sets of parameters of male tenure. The upper curve represents a run beginning from $\bar{r}_{ij} = 1$ (all individuals perfectly related), and the lower curve is from $\bar{r}_{ij} = 0$ (all individuals unrelated). Convergence is rapid, and the calculations were always terminated after 500 epochs. Replicates produced very similar results (see table E).

of r_{ij}, either $r_{ii} = 1$, $r_{ij} = 0$ ($\bar{r} = 0$, females entirely unrelated), or $r_{ij} = 1$ ($\bar{r} = 1$, females perfectly related). Since the average generation time in this model (given the fertility schedule in equations 1) is 9.5 years or 4.75 epochs, the 500 epochs for which each set of calculations was continued is equivalent to 1,000 years or more than 100 generations.

Eight different sets of initial conditions and parameters of male reproductive success were used, and each of these was twice replicated. Table E gives the estimates derived from each set of conditions for $E(r_{ij})$, its standard deviation across time, and the average standard deviation of r_{ij} within the troop at any given time.

Figure B shows that $E(r_{ij})$ varies considerably as a function of the difference in age (D) between two females and that this pattern is affected by the standard deviation of male tenure. The especially low $\bar{r}(D)$ between two females who are separated by four years in age is presumably due to the fact that they are less likely than are adjacent females to share a common father, but cannot be mother and daughter. Two females separated by six years can be mother and daughter, but cannot have the same father. The youngest and

TABLE E. Summary of numerical results. Four distributions of male tenure and two initial distributions of female relatedness were employed. $\bar{T}$ is average male tenure in years, and σ_T is its standard deviation. Entries under $\bar{\bar{r}}_0$ are final cumulative average r_{ij} for runs beginning from populations of unrelated females ($r_{ij} = 0$), while those under $\bar{\bar{r}}_1$ are for initial populations of perfectly related females ($r_{ij} = 1$). $E(r_{ij})$ is the average of the four $\bar{r}$ corresponding to a given distribution of male tenure; it is the value toward which the r appear to be converging. The average sample standard deviation of within-troop r_{ij} (s_I) is approximately 0.16 in all cases, indicating the wide range of relationships that exists even in a troop with only eight adult members. The variability of $\bar{r}$ across epochs is low, however, as indicated by $s_E \approx .03$.

$\bar{T}$	σ_T	$\bar{\bar{r}}_0$	$\bar{\bar{r}}_1$	$E(r_{ij})$	s_I	s_E
3	1	0.120	0.137	0.132	0.162	0.030
		.127	.142			
4	0	.147	.154	.149	.163	.028
		.144	.150			
4	1	.155	.161	.160	.160	.031
		.158	.165			
4	2	.168	.177	.174	.159	.029
		.170	.179			

oldest females have lower average coefficients of relationship to their troopmates, or $\bar{r}(i)$, than do middle-aged females, but the difference is slight.

Figure C illustrates the strong and highly significant positive correlation that exists between the standard deviation of male tenure and r.

Discussion   These results suggest that in troops structured as those near Mount Abu seem to be, adult females are related on average by at least 1/8 (0.125), and quite possibly by 1/7 (0.143) to 1/6 (0.167). But these estimates may be low, because the model used to derive them does not account for several processes, each of which would tend to raise $E(r_{ij})$. One of these processes is inbreeding. If males ever remain with a troop long enough to mate with their daughters, and are accepted by them, resulting offspring would be especially highly related to their mothers, and the offspring of such an inbred individual would be more highly related among them-

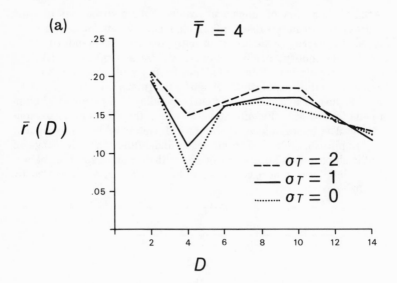

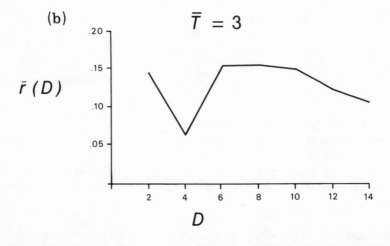

Figure B. Expected r_{ij} as a function of difference in age between females. Individuals four years apart may be typically only half as related to each other as they are on average to their other troopmates, given the assumptions of the model. (a) Increasing the variance of male tenure partly erases this anomaly because a greater proportion of males father three successive offspring, all of whom are therefore at least half siblings ($r_{ij} = 0.25$). (b) Where the average male tenure is three years each male fathers one or two offspring, never three. As a result, individuals four years apart are never siblings of any kind, and show an average r_{ij} of about 1/16, equivalent to that of half first cousins.

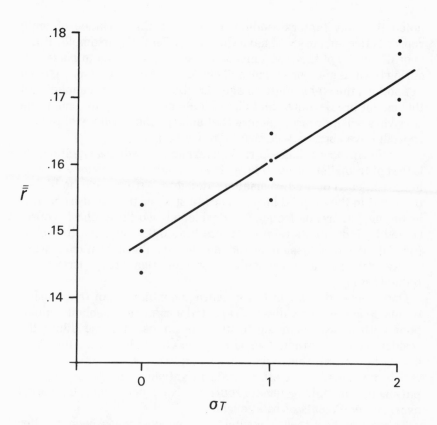

Figure C. Correlation between standard deviation of male tenure and average relatedness among Females, for $\bar{T}=4$. At each σ_T, the upper two points are from replicates beginning at $r_{ij}=1$, and the lower two are from corresponding replicates beginning at $r_{ij}=0$ (see table E). In all cases $\bar{\bar{r}}$ is converging toward an $E(r_{ij})$ very close to the least-squares regression line. The correlation for these sample points 0.93, $P < 0.001$.

selves than are the equivalent offspring of an outbred individual. Even if daughters somehow know who their fathers are and refuse to mate with them, inbreeding will result if successive adult males are sometimes related to each other. Until detailed observational or biochemical genetic studies of the large-scale population structure are carried out for langurs, there will be no way to estimate the magnitude of this "background" inbreeding.

Figures B and C show how increased variance of male reproductive success, caused here by variance of tenure, leads to higher average coefficients of relationship among females. It should be noted that *any* factors tending to increase the variance of male reproductive success will have the same effect. For example, if the overall quality of the environment varies in time so as to make off-spring born at one period more likely to survive than offspring born at another, the stable uniform age distribution used for simplicity in this model breaks down and coefficients of relationship among the survivors will increase. Factors that enlarge the variance of female reproductive success will also increase $E(r_{ij})$.

Another process likely to raise average relatedness significantly is that of troop fissioning. If large troops tend to divide along kinship lines, r in each of the two smaller "daughter" troops will be higher than that in the original large troop. In spite of the small troop size in the model, troops frequently developed two "branches" repre-sented by individuals related at less than 0.001 between branches (fig. D). Because fissioning did not occur, these branches were always rejoined later via males, but their temporary persistence reduced $E(r_{ij})$.

One process that might *lower* average coefficients of relationship is outside mating. Females are reported sometimes to solicit copula-tions with males belonging to all-male bands. If these dilute the resident male's reproductive success (in effect shortening the aver-age male tenure), they lower $E(r_{ij})$.

It is not clear what effect male infanticide would have on the parameters of male or female reproductive success, and thence on average coefficients of relationship.

Taken together, these considerations suggest that $\bar{r}$ probably lies somewhere between 1/8 and 1/4 among adult females in typical troops of the Mount Abu region. As table E shows, for a given mating system this average is reasonably stable across time. How-ever, at any given time there is obviously considerable variance of r_{ij} *within* the troop (standard deviation approximately equal to the mean). If individuals know directly who are their closest kin (for instance, if females know their own adult offspring), selection may often favor particular acts conditioned on estimates of r_{ij} other than $\bar{r}$.

I thank Frank Benford, Department of Engineering and Applied Physics, Harvard University, for several helpful suggestions.

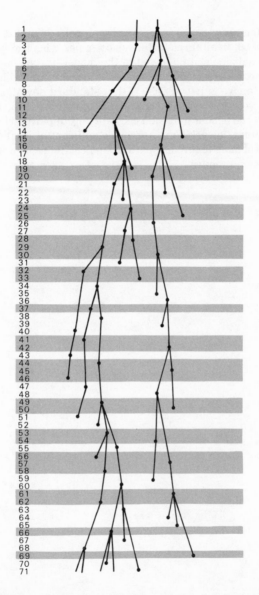

Figure D. Representative family tree, illustrating the development of extended parallel matrilines, and the effect of the variance of male tenure on the frequency of links between them. The diagram shows descent through females for seventy-one epochs near the end of a run. Alternating light and dark bands cover periods of tenure by successive unrelated males. Thus the females born at epochs 3, 4, and 5 have a common father, as do the females at 6 and 7. The tree's two continuous branches, which derive from a single female ancestor, are linked at epochs 28 and 29, at 34 and 36, and at 57 and 58, in all three instances by males who remained with the troop for the maximum span allowed in the model (three epochs). Each of these links raises the average relatedness between the matrilines. In this run the average tenure is two epochs with a standard deviation of one; when the average was two with a standard deviation of zero, no males ever remained three epochs and such cross-linking of parallel female lineages was less frequent, lowering the average coefficient of relationship. Hence, the variance of male tenure affects average relatedness in the troop independently of mean tenure.

Appendix 4 / Average Male Tenure at Abu

Diagram showing nine male take-overs in five langur troops at Abu during 248 months that the histories of those troops were followed. This diagram provides the basis for an estimate of at least one take-over every 27.6 months.[a]

First month of study period	Hillside	Bazaar	Toad Rock	School	IPS
June 1971	Mug 1. Shifty (August 1971)	3 males 2. Shifty (May 1972)	Splitear	Harelip	LeGrand
June 1972	3. Mug				
February 1973			4. Toad		
December 1973	5. Righty				6. (Fission)
April 1975		7. Mug 8. Righty (May 1975)			/ \ new male new male
October 1975[b]	9. Slash-neck				
Number of months history of troop was followed	52	52	48	48	48

a. Date of take-over is given when known.
b. Reported by J. Malcolm.

327

Additions to the Bibliography, 1980

Angst, W., and D. Thommen
1977 New data and a discussion of infant-killing in Old World monkeys and apes. *Folia Primatologica* 27: 198-229.
Blaffer Hrdy, S.
1979 Infanticide among animals: A review, classification, and examination of the implications for the reproductive strategies of females. *Journal of Ethology and Sociobiology* 1: 13-40.
Boggess, J.
1979 Troop male membership changes and infant killing in langurs (*Presbytis entellus*). *Folia Primatologica* 32: 65-107.
Butynski, T.
1980 Infanticide in the blue monkey, *Cercopithecus mitis stuhlmanni*, in the Kibale Forest. Paper prepared for presentation at the Eighth Congress of the International Primatological Society, Florence, Italy.
Curtin, R., and P. Dolhinow
1978 Primate social behavior in a changing world. *American Scientist* 66: 468-475.
1979 Infanticide among langurs—a solution to overcrowding? *Science Today* 13: 35-41.
Galat-Luong, A., and G. Galat
1979 Consequences comportementales de perturbations sociales repetées sur une troupe de mones de Lowe *Cercopithecus cambelli lowei* de Côte-D'Ivoire. *La terre et la vie* 33: 49-57.
Goodall, J.
1977 Infant-killing and cannibalism among free-living chimpanzees. *Folia Primatologica* 28: 259-282.
Labov, J.
1980 Factors affecting infanticidal behavior in wild male house mice (*Mus musculus*). *Journal of Behavioral Ecology and Sociobiology* 6: 297-303.
Mallory, F. F., and R. J. Brooks
1978 Infanticide and other reproductive strategies in the collared lemming *Dicrostonyx groenlandicus*. *Nature* 273: 144-146.
Rudran, R.
1979 Infanticide in red howlers (*Alouatta seniculus*) of northern Venezuela. Paper presented at the Seventh International Congress of Primatology, Bangalore, India.
Vogel, C.
1979 Der Hanuman-Langur (*Presbytis entellus*), ein Parade-Exempel für die theoretischen Konzepte der "Soziobiologie"? *Verhandlungen der Deutschen Zoologischen Gesellschaft* 1979: 73-89.
Wolf, K.
1980 Social change and male reproductive strategy in silvered leaf-monkeys, *Presbytis cristata*, in Kuala Selangor, Peninsular Malaysia. Abstract. *American Journal of Physical Anthropology* 52: 294.

Bibliography

Aldrich-Blake, P.
1970 Problems of social structure in forest monkeys. In *Social Behavior in Birds and Mammals*. New York: Academic Press.

Alexander, B. K.
1970 Parental behavior of adult male Japanese monkeys. *Behavior* 36: 270-285.

Alexander, B. K., and J. H. Bowers
1969 Social organization of a troop of Japanese monkeys in a two-acre enclosure. *Folia Primatologica* 10: 230-242.

Alexander, B. K., and J. Hughes
1971 Canine teeth and rank in Japanese monkeys (*Macaca fuscata*). *Primates* 12: 91-93.

Allen, G. M.
1938 *The Mammals of China and Mongolia*, part 1. New York: American Museum of Natural History Press.

Altmann, J.
1974 Observational study of behavior: sampling methods. *Behavior* 48: 227-265.

Altmann, S., J. Altmann, G. Hausfater, and S. A. McCuskey
In press Life history of yellow baboons: physical development, reproductive parameters and infant mortality. *Primates* 18.

Ayer, A. A.
1948 *The anatomy of* Semnopithecus entellus. Madras: Indian Publishing House.

Badham, M.
1967 A note on the breeding of the spectacled leaf monkey. *International Zoo Yearbook* 7: 89.

Baird, Sir Dugald
1945 Social and economic factors on stillbirths and neonatal deaths. *Journal of Obstetrics and Gynaecology of the British Empire* 52: 217-234.

Baldwin, L. A., M. Kavanagh, and G. Teleki
1975 Field research on langur and proboscis monkeys: an historical, geographic, and bibliographical listing. *Primates* 16: 351-363.

Balikci, A.
1970 *The Netsilik Eskimo.* New York: Natural History Press.

Bardwick, J. M.
1974 The sex hormones, the central nervous system and affect variability in humans. In *Women in Therapy*, ed. V. Franks and V. Burtle. New York: Brunner-Mazel.

Barnicot, N. A., and D. Hewett-Emmett
1972 Red cell and serum proteins of *Cercocebus, Presbytis, Colobus,* and certain other species. *Folia Primatologica* 17: 442-457.

Bauchop, T., and R. W. Martucci
1968 Ruminant-like digestion of the langur monkey. *Science* 161: 698-700.

Beck, B., and R. Tuttle
1972 The behaviour of gray langurs at a Ceylonese waterhole. In *The Functional and Evolutionary Biology of the Primates*, ed. R. Tuttle. Chicago: Aldine.

Ben Shaul, D. M.
1962a Notes on hand-rearing various species of mammals. *International Zoo Yearbook* 4: 300-332.
1962b The composition of the milk of wild animals. *International Zoo Yearbook* 4: 333-342.

Bernstein, I. S.
1968 The lutong of Kuala Selangor. *Behavior* 32: 1-15.
1969a Stability of the status hierarchy in a pigtail monkey group (*Macaca nemestrina*). *Animal Behavior* 17: 452-458.
1969b Spontaneous reorganization of a pigtail monkey group. *Proceedings of the Second International Congress of Primatology* 1: 48-51.
1970 Primate status hierarchies. In *Primate Behavior: Developments in Field and Laboratory Research*, vol. 1, ed. L. A. Rosenblum. New York: Academic Press.
1974 Aggression and social controls in rhesus monkey (*Macaca mulatta*) groups revealed in group formation studies. *Folia Primatologica* 21: 81-107.

Bernstein, I. S., and L. G. Sharpe
1966 Social roles in a rhesus monkey group. *Behavior* 26: 91-104.

Bertram, B.
1976 Kin selection in lions and in evolution. In *Growing Points in Ethol-*

ogy, ed. P. P. G. Bateson and R. A. Hinde. Cambridge: Cambridge University Press.

Biocca, E.

1971   Yanoáma. New York: Dutton.

Bishop, N.

1975a   Vocal behavior of adult male langurs in a high altitude environment. Paper presented at the Forty-fourth Annual Meeting of the American Association of Physical Anthropologists.

1975b   Social behavior of langur monkeys (Presbytis entellus) in a high altitude environment. Ph.D. diss., University of California, Berkeley.

In press   Langurs at high altitudes. Journal of the Bombay Natural History Society.

Blaffer Hrdy, S.

1974   Male-male competition and infanticide among the langurs (Presbytis entellus) of Abu, Rajasthan. Folia Primatologica 22: 19-58.

1975   Male and female strategies of reproduction among the langurs of Abu. Ph.D. diss., Harvard University.

1976   The care and exploitation of nonhuman primate infants by conspecifics other than the mother. Advances in the Study of Behavior 6: 101-158.

1977a   Infanticide as a primate reproductive strategy. American Scientist 65: 40-49.

1977b   Allomaternal care and abuse of infants among hanuman langurs. Sixth Congress of the International Primatological Society, New York: Academic Press.

Blaffer Hrdy, S., and D. B. Hrdy

1976   Hierarchical relations among female Hanuman langurs (Primates: Colobinae: Presbytis entellus). Science 193: 913-915.

Blurton Jones, N.

1972   Comparative aspects of mother-child contact. In Ethological Studies of Child Behaviour, ed. N. Blurton Jones. London: Cambridge University Press.

Boggess, J.

1976   The social behavior of the Himalayan langur (Presbytis entellus) in Eastern Nepal. Ph.D. diss., University of California, Berkeley.

Boelkins, R. C.

1962   Large-scale rearing of infant rhesus monkeys (M. mulatta) in the laboratory. International Zoo Yearbook 4: 285-289.

Bonte, M., and H. von Balen

1969   Prolonged lactation and family spacing in Ruanda. Journal of Biosocial Science 1:97.

Booth, A. H.

1957   Observations on the natural history of the olive colobus monkey, Procolobus verus (van Beneden). Proceedings of the Zoological Society of London 129: 421-431.

Bourlière, F.

1954 *The Natural History of Mammals.* New York: Knopf.

Bourlière, F., C. Hunkeler, and M. Bertrand
1970 Ecology and behavior of Lowe's guenon (*Cercopithecus campbelli lowei*) in the Ivory Coast. In *Old World Monkeys*, ed. J. R. Napier and P. H. Napier. New York: Academic Press.

Bowden, D., P. Winter, and D. Ploog
1967 Pregnancy and delivery behavior in the squirrel monkey (*Saimiri scireus*) and other primates. *Folia Primatologica* 5: 1-42.

Brander, A. A. D.
1939 Behavior of monkeys when attacked. *Journal of the Bombay Natural History Society* 41: 165.

Bruce, H. M.
1960 A block to pregnancy in the house mouse caused by the proximity of strange males. *Journal of Reproduction and Fertility* 1: 96-103.

Burt, W. H.
1943 Territoriality and home range concepts as applied to mammals. *Journal of Mammalogy* 24: 346-352.

Burton, F. D.
1972 The integration of biology and behavior in the socialization of *Macaca sylvana* of Gibraltar. In *Primate Socialization*, ed. F. Poirier. New York: Random House.

Bygott, J. D.
1972 Cannibalism among wild chimpanzees. *Nature* (London) 238: 410-411.

Calhoun, J. B.
1962 Population density and social pathology. *Scientific American* 206: 139-148.

Carpenter, C. R.
1942 Societies of monkeys and apes. *Biological Symposium* 8: 177-204.

Chagnon, N. A.
1968 *Yanomamö, the Fierce People.* New York: Holt, Rinehart and Winston.
1972 Tribal social organization and genetic microdifferentiation. In *The Structure of Human Populations*, ed. G. A. Harrison and A. J. Boyce. Oxford: Clarendon Press.

Chalmers, N. R.
1972 Comparative aspects of early infant development in some captive cercopithecines. In *Primate Socialization*, ed. F. Poirier. New York: Random House, pp. 63-82.

Clutton-Brock, T. H.
1974 Primate social organization and ecology. *Nature* (London) 250: 539-542.
1975 Ranging behaviour of red colobus (*Colobus badius tephrosceles*) in the Gombe National Park. *Animal Behavior* 23: 706-722.

Collias, N., and C. H. Southwick
1952 A field study of population density and social organization in

howling monkeys. *Proceedings of the American Philosophical Society* 96: 143-156.

Conaway, C. H., and C. B. Koford
1965 Estrous cycles and mating behavior in a free-ranging band of rhesus monkeys. *Journal of Mammalogy* 45: 577-588.

Crawford, M. P.
1940 The relation between social dominance and the menstrual cycle in female chimpanzees. *Journal of Comparative Psychology* 30: 483-513.
1942 Dominance and social behavior for chimpanzees in a non-competitive situation. *Journal of Comparative Psychology* 33: 267-277.

Crook, J. H.
1970 The socio-ecology of primates. In *Social Behaviour in Birds and Mammals*, ed. J. H. Crook. New York: Academic Press.
1971 Sources of cooperation in animals and man. In *Man and Beast: Comparative Social Behavior*, ed. J. F. Eisenberg and W. S. Dillon. Washington, D. C.: Smithsonian Institution Press.

Curtin, R. A.
1975 The socioecology of the common langur, *Presbytis entellus*, in the Nepal Himalaya. Ph.D. diss., University of California, Berkeley.
1977 Langur social behavior and infant mortality. *Kroeber Anthrological Society Papers* 50: 27-36.

Darling, F. F.
1937 *A Herd of Red Deer*. London: Oxford University Press.

Darwin, C.
1859 (1967 facsimile) *On the Origin of Species*. New York: Atheneum.

David, G. F. X., and L. S. Ramaswami
1969 Studies on menstrual cycles and other related phenomena in the langur (*Presbytis entellus entellus*). *Folia Primatologica* 11: 300-316.

Delson, E.
1973 Fossil colobine monkeys of the circum-Mediterranean region and the evolutionary history of the Cercopithecidae (Primates, Mammalia). Ph.D. diss., Columbia University.

DeVore, I.
1962 The social behavior and organization of baboon troops. Ph.D. diss., University of Chicago.
1963 Mother-infant relations in free-ranging baboons. In *Maternal Behavior in Mammals*, ed. H. L. Rheingold. New York: Wiley.
1965 Male dominance and mating behavior in baboons. In *Sex and Behavior*, ed. F. Beach. New York: Wiley.

Dickeman, M.
1975 Demographic consequences of infanticide in man. *Annual Review of Ecology and Systematics* 6: 107-137.

Dolhinow, P.
1972 The north Indian langur. In *Primate Patterns*, ed. P. Dolhinow. New York: Holt, Rinehart and Winston.
1977 Letter to the editors. *American Scientist* 65: 266.

Dolhinow, P. J., and N. Bishop
1972 The development of motor skills and social relationships among primates through play. In *Primate Patterns,* ed. P. Dolhinow. New York: Holt, Rinehart and Winston.

Douglas, J. W. B.
1946 The extent of breast feeding in Great Britain in 1946. *Journal of Obstetrics and Gynaecology of the British Empire* 57: 343.

Downhower, J. F., and K. B. Armitage
1971 The yellow-bellied marmot and the evolution of polygamy. *American Naturalist* 105: 355-370.

Drickamer, L.
1974 A ten-year summary of reproductive data for free-ranging *Macaca mulatta. Folia Primatologica* 21: 61-80.

Drillien, C., and F. Richmond
1956 Prematurity in Edinburgh. *Archives of Disease in Childhood* 31: 390-394.

DuMond, F. V.
1968 The squirrel monkey in a semi-natural environment. In *The Squirrel Monkey,* ed. L. Rosenblum and R. W. Cooper. New York: Academic Press.

Dunbar, R. I. M., and E. P. Dunbar
1974a Behaviour related to birth in wild gelada baboons (*Theropithecus gelada*). *Behaviour* 50: 185-191.
1974b Ecology and population genetics of *Colobus guereza* in Ethiopia. *Folia Primatologica* 21: 188-208.

Ehrman, L.
1972 Genetics and sexual selection. In *Sexual Selection and the Descent of Man 1871-1971,* ed. B. G. Campbell. Chicago: Aldine.

Eisenberg, J. F., N. A. Muckenhirn, and R. Rudran
1972 The relation between ecology and social structure in primates. *Science* 176: 863.

Ellefson, J. O.
1968 Territorial behavior in the common white-handed gibbon *Hylobates lar* Linn. In *Primates,* ed. P. Jay. New York: Holt, Rinehart and Winston.

Ellerman, J. R., and T. C. S. Morrison-Scott
1951 *Checklist of Palaearctic and Indian Mammals 1758-1946.* London: British Museum (Natural History).

Fisher, R. A.
1958 *The Genetical Theory of Natural Selection.* New York: Dover.

Fossey, D.
1974 The behavior of the great apes. Paper prepared for participation in Burg Wartenstein Symposium No. 62, July 1974.
1976 The behaviour of the mountain gorilla. Ph.D. diss., Cambridge University.

Furuya, Y.
1962   Social life of silvered leaf monkeys. *Primates* 3: 41-60.

G., J. F.
1902   Habits of the langoor monkey. *Journal of the Bombay Natural History Society* 14: 149-151.

Gadgil, M., and W. H. Bossert
1970   Life historical consequences of natural selection. *American Naturalist* 104: 1-24.

Gartlan, J. S.
1969   Sexual and maternal behavior of the vervet monkey, *Cercopithecus aethiops. Journal of Reproduction and Fertility,* supplement 6: 137-150.

Gee, E. P.
1955   A new species of langur in Assam. *Journal of the Bombay Natural History Society* 53: 252-254.
1961   The distribution and feeding habits of the golden langur, *Presbytis geei* Gee (Khajura, 1956). *Journal of the Bombay Natural History Society* 58: 1-12.

Geist, V.
1971   *Mountain Sheep.* Chicago: University of Chicago Press.

Gillman, J., and C. Gilbert
1946   The reproductive cycles of the chacma baboon with special reference to the problems of menstrual irregularities as assessed by the behavior of the sex skin. *South African Journal of Medical Science (Biology Supplement)* 11: 1-54.

Glander, K. E.
1974   Baby-sitting, infant sharing, and adoptive behavior in mantled howling monkeys. Abstract of paper presented at the Forty-third Annual Meeting of the American Association of Physical Anthropologists. *American Journal of Physical Anthropology* 41: 482.

Gomila, J.
1975   Fertility differentials and their significance for human evolution. In *The Role of Natural Selection in Human Evolution,* ed. F. M. Salzano. North Holland Publishing Company.

Goodall, J.
See Lawick-Goodall, J. van.

Gouzoules, H.
1974   Harassment of sexual behaviour in the stumptail macaque (*Macaca arctoides*). *Folia Primatologica* 17: 1-19.

Groves, C. P.
1970   The forgotten leaf-eaters, and the phylogeny of the Colobinae. In *Old World Monkeys,* ed. J. R. Napier and P. H. Napier. New York: Academic Press.
1973   Notes on the ecology and behavior of the Angola colobus (*Colobus angolensis* P. L. Sclater, 1860) in N. E. Tanzania. *Folia Primatologica* 20: 12-26.

Hall, K. R. L.
1968 Behavior and ecology of the wild patas monkey (*Erythrocebus patas*) in Uganda. In *Primates*, ed. P. C. Jay. New York: Holt, Rinehart and Winston.
Hall, K. R. L., and I. DeVore
1965 Baboon social behavior. In *Primate Behavior*, ed. I. DeVore. New York: Holt, Rinehart and Winston.
Hamburg, D. A.
1969 Observations of mother-infant interactions in primate field studies. In *Determinants of Infant Behavior*, vol. 4, ed. B. M. Foss. London: Methuen.
Hamilton, W. D.
1963 The evolution of altruistic behaviour. *American Naturalist* 97: 354-356.
1964 The genetical evolution of social behavior, parts 1 and 2. *Journal of Theoretical Biology* 7: 1-51.
1971 Selection of selfish and altruistic behavior in some extreme models. In *Man and Beast: Comparative Social Behavior*, ed. J. P. Eisenberg and W. S. Dillon. Washington, D. C.: Smithsonian Institution Press.
1972 Altruism and related phenomena mainly in social insects. *Annual Review of Ecology and Systematics* 3: 193-232.
Hamilton, W. J., III
1973 *Life's Color Code*. New York: McGraw-Hill.
Hanby, J. P., L. T. Robertson, C. H. Phoenix
1971 The sexual behavior of a confined troop of Japanese macaques. *Folia Primatologica* 16: 123-143.
Harlow, H. F.
1971 *Learning to Love*. New York: Ballantine Books.
Harlow, H. F., M. K. Harlow, R. O. Dodsworth, and G. L. Arling
1966 Maternal behavior of rhesus monkeys deprived of mothering and peer associations in infancy. *Proceedings of the American Philosophical Society* 110: 58-66.
Harrison, T.
1962 Leaf monkeys at Fraser's Hill. *Malayan Nature Journal* 16: 120-125.
Hartman, C. G.
1932 Studies in the reproduction of the monkey *Macaca* (*Pithecus*) *rhesus*, with special reference to lactation and pregnancy. *Contributions to Embryology* 134: 3-160.
Hausfater, G.
1974 Dominance and reproduction in baboons (*Papio cynocephalus*): a quantitative analysis. Ph.D. diss., University of Chicago.
Hess, J. P.
1973 Some observations on the sexual behaviour of captive lowland gorillas *Gorilla G. gorilla* (Savage and Wyman). In *Comparative Ecology*

and Behaviour of Primates, ed. R. P. Michael and J. H. Crook. New York: Academic Press.

Hill, C. A.
1972   Infant-sharing in the family Colobidae emphasizing Pygathrix. Primates 13: 195-200.
1975   The longevity record for Colobus. Primates 16: 235.

Hill, W. C. O.
1934   A monograph on the purple-faced leaf-monkeys (Pithecus vetulus). Ceylon Journal of Science, section B 1, vol. 9: 23-88.
1936   Supplementary observations on purple-faced leaf-monkeys (Genus Kasi). Spolia Zeylanica 20: 115-133.
1939   An annotated systematic list of the leaf-monkeys. Ceylon Journal of Science 21: 277-305.

Hinde, R. A.
1970   Animal Behaviour: A Synthesis of Ethology and Comparative Psychology. New York: McGraw-Hill.
1975   The concept of function. In Function and Evolution in Behaviour, ed. G. Baerends, C. Beer, and A. Manning. Oxford: Clarendon Press.

Hinde, R., and Y. Spencer-Booth
1967   The behaviour of socially living rhesus monkeys in their first two and a half years. Animals Behaviour 15: 169-196.
1969   The effect of social companions on mother-infant relations in rhesus monkeys. In Primate Ethology, ed. D. Morris. New York: Anchor.

Hingston, R. W. G.
1920   A Naturalist in Himalaya. London: Witherby.

Hirshfield, M. F., and D. W. Tinkle
1975   Natural selection and the evolution of reproductive effort. Proceedings National Academy Science 72: 2227-2231.

Hladik, C. M. and A. Hladik
1972   Disponibilités alimentaires et domaines vitaux des primates a Ceylan. La Terre et la vie 26: 149-215.

Hooijer, D. A.
1962   Quarternary langurs and macaques from the Malay Archipelago. Zoologische verhandelingen (Leiden) 55: 1-64.

Horwich, R. H.
1974   Development of behaviors in a male spectacled langur (Presbytis obscurus). Primates 15: 151-178.

Horwich, R., and D. Manski
1975   Maternal care and infant transfer in two species of Colobus monkeys. Primates 16: 49-73.

Hrdy, D. B., N. A. Barnicot, C. A. Alper
1975   Protein polymorphism in the hanuman langur (Presbytis entellus). Folia Primatologica 24: 173-187.

Hughes, T. H.
1884   An incident in the habits of Semnopithecus entellus, the common

hanuman monkey. *Proceedings of the Asiatic Society of Bengal*, pp. 147-150.

Itani, J.
1959 Paternal care in the wild Japanese monkey, *Macaca fuscata fuscata*. *Primates* 2: 61-93.
Iwamoto, T.
1974 A bioeconomic study of a provisioned troop of Japanese monkeys (*Macaca fuscata fuscata*) at Koshima Islet, Miyazaki. *Primates* 15: 241-262.

Jay, P. C.
1962 Aspects of maternal behavior among langurs. *Annals of the New York Academy of Sciences* 102: 468-476.
1963a The Indian langur monkey (*Presbytis entellus*). In *Social Behavior*, ed. C. H. Southwick. Princeton: Van Nostrand.
1963b Mother-infant relations in free-ranging langurs. In *Maternal Behavior in Mammals*, ed. H. L. Rheingold. New York: Wiley.
1963c The social behavior of the langur monkey. Ph.D. diss., University of Chicago.
1965 The common langur of North India. In *Primate Behavior*, ed. I. DeVore. New York: Holt, Rinehart and Winston.
Jolly, A.
1972a *The Evolution of Primate Behavior*. New York: Macmillan.
1972b Hour of birth in primates and man. *Folia Primatologica* 18: 108-121.

Kaufmann, J. H.
1962 Ecology and social behavior of the coati, *Nasua narica*, on Barro Colorado Island, Panama. *University of California Publications in Zoology* 60: 95-222.
1965 A three-year study of mating behaviour in a free-ranging band of monkeys. *Ecology* 46: 500-512.
Kavanagh, M.
In press The social behaviour of Doucs (*Pygathrix nemaeus nemaeus*).
Kawabe, M., and T. Mano
1972 Ecology and behavior of the wild proboscis monkey, *Nasalis larvatus* (Wurmb), in Sabah, Malaysia. *Primates* 13: 213-228.
Kawai, M.
1965a (first published 1958) On the system of social ranks in a natural troop of Japanese monkeys, part 1. In *Japanese Monkeys: A Collection of Translations*, ed. S. Altmann. University of Alberta: S. Altmann.
1965b (first published 1958) On the system of social ranks in a natural troop of Japanese monkeys, part 2. In *Japanese Monkeys: A Collection of Translations*, ed. S. Altmann. University of Alberta: S. Altmann.
Kern, J. A.

1964   Observations on the habits of the proboscis monkey, *Nasalis larvatus* WURMB, made in the Brunei Bay area, Borneo. *Zoologica* (New York) 49: 183-192.

King, J. A.

1973   The ecology of aggressive behavior. *Annual Review of Ecology and Systematics* 4: 117-138.

Koford, C. B.

1963   Ranks of mothers and sons in bands of rhesus monkeys. *Science* 141: 356-357.

1966   Population changes in rhesus monkeys 1960-1965. *Tulane Studies in Zoology* 13: 1-7.

Konner, M. J.

1972   Aspects of the developmental ethology of a foraging people. In *Ethological Studies of Child Behavior*, ed. N. Blurton Jones. Cambridge: Cambridge University Press.

Kruuk, H.

1972   *The Spotted Hyena*. Chicago: University of Chicago Press.

Kuhl, E.

1925   *Studies in Philology*. Chapel Hill: University of North Carolina.

Kummer. H.

1967   Tripartite relations in Hamadryas baboons. In *Social Communication among Primates*, ed. S. Altmann. Chicago: University of Chicago Press.

1968   *Social Organization of Hamadryas Baboons*. Chicago: University of Chicago Press.

Kummer, H., W. Gotz, and W. Angst

1974   Triadic differentiation: an inhibitory process protecting pair bonds in baboons. *Behaviour* 49: 62.

Lack, D.

1954   The evolution of reproductive rates. In *Evolution as a Process*, ed. J. S. Huxley, A. C. Hardy, and E. B. Ford. London: Allen and Unwin.

Lancaster, J.

1971   Play-mothering: the relations between juvenile females and young infants among free-ranging vervet monkeys (*Cercopithecus aethiops*). *Folia Primatologica* 15: 161-182.

Langer, W.

1972   Checks on population growth: 1750-1850. *Scientific American* 226: 92-99.

Lasley, J. F., and R. Bogart

1943   Some factors influencing reproductive efficiency of range cattle under artificial and natural breeding conditions. *Missouri Agricultural Exposition Station Research Bulletin 376*.

Laughlin, W. S.

1968   The demography of hunters: an Eskimo example, part 4: Discussions. In *Man the Hunter*, ed. R. B. Lee and I. DeVore. Chicago: Aldine.

Lawick, H. van
1973 *Solo: The Story of an African Wild Dog Puppy and Her Pack.* London: Collins.
Lawick, H. van, and J. van Lawick-Goodall
1971 *Innocent Killers.* Boston: Houghton-Mifflin.
Lawick-Goodall, J. van
1967 Mother-offspring relationships in free-ranging chimpanzees. In *Primate Ethology*, ed. D. Morris. Chicago: Aldine.
1968 The behaviour of free-living chimpanzees in the Gombe Stream Reserve. *Animal Behaviour Monograph* 1: 165-311.
1971 *In the Shadow of Man.* Boston: Houghton-Mifflin.
LeBoeuf, B. J., R. J. Whiting, and R. F. Gantt
1972 Perinatal behavior of northern elephant seal females and their young. *Behaviour* 43: 121-156.
Lee, R. B., and I. DeVore, eds.
1968 *Man the Hunter.* Chicago: Aldine.
Lehrman, D. S.
1961 Hormonal regulation of parental behavior in birds and infrahuman mammals. In *Sex and Internal Secretion*, ed. W. C. Young. Baltimore: Williams and Wilkins.
Leskes, A., and N. H. Acheson
1971 Social organization of a free-ranging troop of black and white colobus monkeys (*Colobus abyssinicus*). *Proceedings of the Third International Congress of Primatology* 3: 22-31.
Levick, G. M.
1915 *Antarctic Penguins: A Study of Their Social Habits.* New York: McBridge, Nast.
Liggins, G. C.
1972 The fetus and birth. In *Embryonic and Fetal Development*, ed. C. R. Austin and R. V. Short. Cambridge: Cambridge University Press.
Lindburg, D. G.
1971 The rhesus monkey in North India: an ecological and behavioral study. In *Primate Behavior*, ed. L. A. Rosenblum. New York: Academic Press.
Lippold, L. K., and D. K. Brockman
1974 San Diego's douc langurs. *Zoonooz* 47: 4-11.
In press The douc langur: a time for conservation. In *Primate Conservation*, ed. Rainer and Bourne. New York: Academic Press.
Lorenz, K.
1966 (1971 edition) *On Aggression.* New York: Bantam.
Lorimer, F.
1954 *Culture and Human Fertility.* Zurich: UNESCO.
Lott, D. F., and J. S. Rosenblatt
1969 Development of maternal responsiveness during pregnancy in the laboratory rat. In *Determinants of Infant Behavior* vol. 4. London: Methuen.

Loy, J.
1970 Peri-menstrual sexual behavior among rhesus monkeys. *Folia Primatologica* 13: 286-297.
1974 Reproduction in patas monkeys: behavioral and physical phenomena related to gestation and parturition. Paper presented at the Forty-Third Annual Meeting of the American Association of Physical Anthropologists.
1975 The descent of dominance in *Macaca*: insights into the structures of human societes. In *Socioecology and Psychology of Primates*, ed. R. Tuttle. The Hague: Mouton.

Marler, P.
1969 *Colobus guereza* territoriality and group composition. *Science* 163: 93-95.
1972 Vocalizations of East African monkeys, part 2: black and white colobus. *Behaviour* 42: 175-197.
Marsden, H. M.
1968 Agonistic behaviour of young rhesus monkeys after changes induced in social rank of their mothers. *Animal Behaviour* 16: 38-44.
Marsden, H. M., and S. H. Vessey
1968 Adoption of an infant green monkey within a social group. *Communications in Behavioral Biology*, part A 6: 275-279.
Martin, R. D.
1975 Ascent of the primates. *Natural History* 84: 53-61.
Martin, M. K., and B. Voorhies
1975 *Female of the Species*. New York: Columbia University Press.
Mason, W. A.
1968 Use of space by *Callicebus* groups. In *Primates*, ed. P. Jay. New York: Holt, Rinehart and Winston.
Maynard Smith, J., and G. R. Price
1973 The logic of animal conflict. *Nature* 246: 15-18.
Mayr, E.
1974 Behavior programs and evolutionary strategies. *American Scientist* 62: 650-659.
McCann, C.
1928 Notes on the common Indian langur (*Pithecus entellus*). *Journal of the Bombay Natural History Society* 33: 192-194.
1933a Notes on the colouration and habits of the white-browed gibbon or hoolock (*Hylobates hoolock* Harl.). *Journal of the Bombay Natural History Society* 36: 395-405.
1933b Observations on some of the Indian langurs. *Journal of the Bombay Natural History Society* 36: 616-628.
1966 *100 Beautiful Trees of India*. Bombay: Taraporevala Sons.
McClintock, M.K.
1971 Menstrual synchrony and suppression. *Nature* 229: 244-245.
1974 Sociobiology and reproduction in the Norway rat (*Rattus norvegi-*

cus): estrous synchrony and the role of the female rat in copulatory behavior. Ph.D. diss., University of Pennsylvania.

McKenna, J. J.

1974a Coming of age in hanuman langur society. Zoonooz 47: 12-18.

1974b Perinatal behavior and parturition of a Colobinae Presbytis entellus entellus (hanuman langur). Laboratory Primate Newsletter 13: 13-15.

Mead, M.

1971 Sex and Temperament in Three Primitive Societies. New York: Dell.

Medway, L.

1970 The monkeys of Sundaland: ecology and systematics of the cercopithecids of a humid equatorial environment. In Old World Monkeys, ed. J. R. Napier and P. H. Napier. New York: Academic Press.

Missakian, E.

1972 Genealogical and cross-genealogical dominance relations in a group of free-ranging rhesus monkeys (Macaca mulatta) on Cayo Santiago. Primates 13: 169-180.

Mitchell, G.

1970 Abnormal behavior in primates. In Primate Behavior, ed. L. Rosenblum. New York: Academic Press.

Mitchell, G., and E. M. Brandt

1972 Paternal behavior in primates. In Primate Socialization, ed. F. Poirier. New York: Random House.

Mizuhara, H.

1946 Social changes of Japanese monkey troops in the Takasakiyama. Primates 5: 27-52.

Mohnot, S. M.

1968 Interactions and social changes in troops of Hanuman langur, Presbytis entellus, in India. Abstracts Symposium of Natural Resources Rajasthan, Jodhpur, October 23-26, 1968.

1971a Ecology and behaviour of the Hanuman langur, Presbytis entellus (Primates: Cercopithecidae) invading fields, gardens, and orchards around Jodhpur, Western India. Tropical Ecology 12: 237-249.

1971b Some aspects of social changes and infant-killing in the hanuman langur, Presbytis entellus (Primates: Cercopithecidae) in Western India. Mammalia 35: 175-198.

1974 Ecology and behavior of the common Indian langur, Presbytis entellus Dufresne. Ph.D. diss., University of Jodhpur.

Mukherjee, R. P., and S. S. Saha

1974 The golden langurs (Presbytis geei Khajuria, 1956) of Assam. Primates 15: 327-340.

Murphy, Y., and R. F. Murphy

1974 Women of the Forest. New York: Columbia University Press.

Napier, J. R., and P. H. Napier

1967   A Handbook of Living Primates. New York: Academic Press.
Neel, J. V.
1970   Lessons from a "primitive" people. Science 170: 818.
Noirot, E.
1972   The onset of maternal behavior in rats, hamsters, and mice: a selective review. Advances in the Study of Behavior 4: 107-146.
Nolte, A.
1955   Field observations on the daily routine and social behavior of common Indian monkeys with special reference to the Bonnet monkey (Macaca radiata Geoffroy). Journal of the Bombay Natural History Society 52: 117-184.

Oppenheimer, J. R.
In Press   Presbytis entellus, the hanuman langur. In Primate Conservation, ed. Rainier and Bourne. New York: Academic Press.
Orians, G. H.
1969   On the evolution of mating systems in birds and mammals. American Naturalist 103: 589-603.

Parker, G. A.
1974   Assessment strategy and the evolution of fighting behaviour. Journal of Theoretical Biology 47: 223-243.
Petter-Rousseaux, A.
1968   Cycles genitaux saisonniers des lemuriens malgaches. In Cycles genitaux saisonniers de mammifères sauvages. Paris: Masson.
Pilgrim, G. E.
1915   New Siwalik primates and their bearing on the question of the evolution of man and the Anthropoidea. Recent Geological Survey of India 45: 1-74.
Pocock, R. I.
1925   The external characteristics of catarrhine monkeys and apes. Proceedings of the Zoological Society of London.
1928a   The langurs, or leaf monkeys, of British India. Journal of the Bombay Natural History Society 32: 472-504.
1928b   The langurs, or leaf monkeys, of British India, part 2. Journal of the Bombay Natural History Society 32: 660-677.
1934   The monkeys of the genera Pithecus (or Presbytis) and Pygathrix found to the east of the Bay of Bengal. Proceedings of the Zoological Society of London.
Poirier, F. E.
1968a   Nilgiri langur (Presbytis johnii) territorial behavior. Primates 9: 351-364.
1968b   The Nilgiri langur (Presbytis johnii) mother-infant dyad. Primates 9: 45-68.
1970   The Nilgiri langur (Presbytis johnii) of South India. In Primate Behavior, vol. 1, ed. L. Rosenblum. New York: Academic Press.

Pournelle, G. H.
1966   Birth of a proboscis monkey. *Zoonooz* 39: 3-7.
Prakash, I.
1962   Group organization, sexual behavior, and breeding season of certain Indian monkeys. *Japanese Journal of Ecology* 12: 83-86.
Pythian-Adams, E. G.
1940   Behavior of monkeys when attacked. *Journal of the Bombay Natural History Society* 41: 653.

Rahaman, H.
1973   The langurs of the Gir sanctuary (Gujarat)—a preliminary survey. *Journal of the Bombay Natural History Society* 70: 295-314.
Rahaman, H., and M. D. Parthasarathy
1962   Studies of the social behavior of bonnet monkeys. *Primates* 10: 149-162.
Ransom, T., and B. Ransom
1971   Adult male infant interactions among baboons (*Papio anubis*). *Folia Primatologica* 16: 179-195.
Ransom, T., and T. Rowell
1972   Early social development of feral baboons. In *Primate Socialization*, ed. F. Poirier. New York: Random House.
Reinhardt, V., and A. Reinhardt
1975   Dynamics of social hierarchy in a dairy herd. *Z. Tierpsychol.* 38: 315-323.
Richards, M. P. M.
1967   Some effects of experience on maternal behavior in rodents. In *Determinants of Infant Behavior*, vol. 4, B. M. Foss. London: Methuen.
Ring, A., and R. Scragg
1973   A demographic and social study of fertility in rural New Guinea. *Journal of Biosocial Science* 5: 89-121.
Ripley, S.
1965   The ecology and social behavior of the Ceylon gray langur, *Presbytis entellus thersites*. Ph.D. diss., University of California, Berkeley.
1967a   The leaping of langurs: problem in the study of locomotor adaptation. *American Journal of Physical Anthropology* 26: 149-170.
1967b   Intertroop encounters among Ceylon gray langurs (*Presbytis entellus*). In *Social Communication among Primates*, ed. S. Altmann. Chicago: University of Chicago Press.
1970   Leaves and leaf monkeys: the social organization of foraging in gray langurs (*Presbytis entellus thersites*). In *Old World Monkeys*, ed. J. R. Napier and P. Napier. New York: Academic Press.
Roberts, C., and C. Lowe
1975   Where have all the conceptions gone? *Lancet* 1: 498-499.
Rodman, P.
1973a   Synecology of Bornean primates, with special reference to the

behavior and ecology of the orang-utan. Ph.D. diss., Harvard University.
1973b Population composition and adaptive organization among orang-utans of the Kutai reserve. In *Comparative Ecology and Behavior of Primates*, ed. R. Michael and J. H. Crook. New York: Academic press.
Rosenblum, L. A.
1968 Mother-infant relations and early behavioral development in the squirrel monkey. In *The Squirrel Monkey*, ed. L. A. Rosenblum and R. W. Cooper. New York: Academic Press.
1971 Infant attachment in monkeys. In *The Origins of Human Social Relations*, ed. R. Schaffer. New York: Academic Press.
1972 Sex and age differences in response to infant squirrel monkeys. *Brain, Behavior, Evolution* 5: 30-40.
Rowell, T. E.
1963 Behaviour and female reproductive cycles of rhesus macaques. *Journal of Reproduction and Fertility* 6: 193-203.
1970 Baboon menstrual cycles affected by social environment. *Journal of Reproduction and Fertility* 21: 133-141.
1972a Female reproduction cycles and social behavior in primates. *Advances in the Study of Behavior*, 4: 69-105.
1972b *The Social Behavior of Monkeys*. Baltimore: Penguin.
1974 The concept of social dominance. *Behavioural Biology* 11: 131-154.
Rowell, T. E., R. A. Hinde, and Y. Spencer-Booth
1964 "Aunt"-infant interaction in captive rhesus monkeys. *Journal of Animal Behaviour* 12: 219-226.
Rudran, R.
1973a The reproductive cycles of two subspecies of purple-faced langurs (*Presbytis senex*) with relation to environmental factors. *Folia Primatologica* 19: 41-60.
1973b Adult male replacement in one-male troops of purple-faced langurs (*Presbytis senex senex*) and its effect on population structure. *Folia Primatologica* 19: 166-192.

Saayman, G. S.
1971 Behaviour of the adult males in a troop of free-ranging chacma baboons (*Papio ursinus*). *Folia Primatologica* 15: 36-37.
Sabater Pi, J.
1973 Contribution to the ecology of *Colobus polykomus satanas* (Waterhouse, 1838) of Rio Muni, Republic of Equatorial Guinea. *Folia Primatologica* 19: 193-207.
Sade, D. S.
1965 Some aspects of parent-offspring and sibling relations in a group of rhesus monkeys, with a discussion of grooming. *American Journal of Physical Anthropology* 23: 1-17.
1967 Determinants of dominance in a group of free-ranging rhesus monkeys. In *Social Communication among Primates*, ed. S. Altmann. Chicago: University of Chicago Press.

1972 A longitudinal study of social behavior of rhesus monkeys. In *The Functional and Evolutionary Biology of the Primates*, ed. R. Tuttle. Chicago: Aldine.

Saylor, A., and M. Salmon
1971 An ethological analysis of communal nursing by the house mouse (*Mus musculus*). *Behaviour* 40: 62-85.

Schaller, G. B.
1963 *The Mountain Gorilla: Ecology and Behavior*. Chicago: University of Chicago Press.
1967 *The Deer and the Tiger*. Chicago: University of Chicago Press.
1972a *The Serengeti Lion: A Study of Predator-Prey Relations*. Chicago: University of Chicago Press.
1972b The sociable kingdom. In *The Marvels of Animal Behavior*, ed. P. Marler. Washington, D. C.: The National Geographic Society.
1973 *Golden Shadows, Flying Hooves*. New York: Knopf.

Schein, M. W., and M. H. Fohrman.
1955 Social dominance relations in a herd of dairy cattle. *British Journal of Animal Behavior* 3: 45-55.

Schenkel, R., and L. Schenkel Hulliger
1966 On the sociology of free-ranging colobus (*Colobus guereza caudatatus* Thomas, 1885). *First Congress of the International Primate Society*. Stuttgart: Fischer.

Scott, J. P., and J. L. Fuller
1965 *Genetics and the Social Behavior of the Dog*. Chicago: University of Chicago Press.

Seay, B.
1966 Maternal behavior in primiparous and multiparous rhesus monkeys. *Folia Primatologica* 4: 146-168.

Simons, E. L.
1972 *Primate Evolution*. New York: Macmillan.

Simpson, M. J. A.
1973 The social grooming of male chimpanzees. In *Comparative Ecology and Behaviour of Primates*, ed. R. P. Michael and J. H. Crook. New York: Academic Press.

Singh, S. D.
1969 Urban monkeys. *Scientific American* 221: 108-115.

Singh, S. D., and S. N. Manocha
1966 Reactions of the rhesus monkey and the langur in novel situations. *Primates* 7: 259-262.

Skutch, A. F.
1961 Helpers among birds. *Condor* 63: 198-226.

Smith, T. E.
1960 The Cocos-Keeling Islands: a demographic laboratory. *Population Studies* 14: 94.

Smuts, B.
1972 Natural selection and macaque social behavior. Senior honors thesis, Harvard University.

Sparks, J.
1969   Allogrooming in primates. In *Primate Ethology*, ed. D. Morris. New York: Doubleday.

Spencer-Booth, Y.
1968   The behavior of group companions towards rhesus monkey infants. *Animal Behaviour* 16: 541-557.
1970   The relationships between mammalian young and conspecifics other than mothers and peers: a review. *Advances in the Study of Behavior* 3: 120-194.

Stehn, R. A., and M. E. Richmond
1975   Male-induced pregnancy termination in the prairie vole, *Microtus ochrogaster*. *Science* 187: 1211-1213.

Stott, K., and G. J. Selsor
1961   Observations of the maroon leaf-monkey in North Borneo. *Mammalia* 25: 184-189.

Struhsaker, T.
1967a   Behavior of vervet monkeys (*Cercopithecus aethiops*). *University of California Publications in Zoology* 82: 1-74.
1967b   Social structure among vervet monkeys (*Cercopithecus aethiops*). *Behavior* 29: 83-121.
1969   Correlates of ecology and social organization among African cercopithecines. *Folia Primatologica* 11: 80-118.
1975   *The Red Colobus Monkey*. Chicago: University of Chicago Press.
1976   Infanticide in the redtailed monkey (*Cercopithecus ascanius*). Paper presented at the Sixth Congress of the International Primatological Society, Cambridge, England.

Struhsaker, T. T., and J. S. Gartlan
1970   Observations on the behaviour and ecology of the patas monkey (*Erythrocebus patas*) in the Waza Reserve, Cameroon. *Journal of Zoology* 161: 49-63.

Sugiyama, Y.
1964   Group composition, population density and some sociological observations of hanuman langurs (*Presbytis entellus*). *Primates* 5: 7-38.
1965a   Behavioral development and social structure in two troops of hanuman langurs (*Presbytis entellus*). *Primates* 6: 213-247.
1965b   On the social change of hanuman langurs (*Presbytis entellus*) in their natural conditions. *Primates* 6: 381-417.
1966   An artificial social change in a hanuman langur troop (*Presbytis entellus*). *Primates* 7: 41-72.
1967   Social organization of hanuman langurs. In *Social Communication among Primates*, ed. S. Altmann. Chicago: University of Chicago Press.
1973   The social structure of wild chimpanzees. In *Comparative Ecology and Behaviour of Primates*, ed. R. P. Michael and J. H. Crook. New York: Academic Press.
1975   Characteristics of the ecology of the Himalayan langurs. *Journal of Human Evolution* 5: 249-277.

1976   Life history of male Japanese monkeys. In *Advances in the Study of Behavior* 7: 255-284.

Sugiyama, Y., K. Yoshiba, and M. D. Parthasarathy
1965   Home range, mating season, male group and intertroop relations in hanuman langurs (*Presbytis entellus*). *Primates* 6: 73-106.

Suzuki, A.
1971   Carnivority and cannibalism observed among forest-living chimpanzees. *Journal of Anthropological Society Nipponese* 79: 30-48.

Tanabe, T. Y., and G. W. Salisbury
1946   Influence of age on breeding efficiency. *Journal of Dairy Science* 29: 337-344.

Tanaka, T., T. Kisaburo, and S. Kotera
1970   Effects on infant loss on the interbirth interval of Japanese monkeys. *Primates* 11: 113-117.

Teleki, G.
1973   The omnivorous chimpanzee. *Scientific American* 228: 33-42.

Terkel, J., and J. S. Rosenblatt
1968   Maternal behavior induced by maternal blood plasma injected into virgin rats. *Journal of Comparative and Physiological Psychology* 65: 479-482.

Thompson, N. S.
1967   Primate infanticide: a note and request for information. *Laboratory Primate Newsletter* 6: 18-19.

Tilson, R., and R. Tenaza
1976   Monogamy and duetting in an Old World monkey. *Nature* 263: 320-321.

Trivers, R. L.
1971   The evolution of reciprocal altruism. *Quarterly Review of Biology* 46: 35-57.
1972   Parental investments and sexual selection. In *Sexual Selection and the Descent of Man 1871-1971*, ed. B. Campbell. Chicago: Aldine.
1974   Parent-offspring conflict. *American Zoologist* 14: 249-264.

Trivers, R. L., and D. Willard
1973   Natural selection of parental ability to vary the sex ratio of offspring. *Science* 179: 90.

Udry, J. R., and N. M. Morris
1968   Distribution of coitus in the menstrual cycle. *Nature* 220: 593-596.

Ullrich, W.
1961   Zur biologie und soziologie der Colobusaffen (*Colobus guereza caudasus* Thomas, 1885). *Zoologische Garten* 25: 305-368.

Vogel, C.
1971   Behavioral differences of *Presbytis entellus* in two different habi-

tats. *Proceedings of the Third International Congress of Primatology* 3: 41-47.

1973a Hanuman as an object for anthropologists—field studies of social behavior among the gray langurs of India. *Contributions to Indian Studies*, vol. 1. Cultural Department, Embassy of the Federal Republic of Germany, New Delhi. Varanasi: Chowkhamba Sanskrit Series Office.

1973b Acoustical communication among free-ranging common Indian langurs (*Presbytis entellus*) in two different habitats of North India. *American Journal of Physical Anthropology* 38: 469-480.

1974 Intergroup relations of *Presbytis entellus* in the Kumaon Hills and in Rajastham (North India). *Proceedings of the Fifth International Congress of Primatology, Nagoya, Japan*. Basel: Karger.

Wagley, C.
1969 Cultural influences on population: a comparison of two Tupi tribes. In *Environment and Cultural Behavior*, ed. A. P. Vayda. New York: Natural History Press.

Wagnon, K. A.
1965 Social dominance in range cows and its effects on supplement feeding. *California Agricultural Exposition Station Bulletin No. 819.*

Warren, J. M.
1967 Discussion of social dynamics. In *Social Communication among Primates*, ed. S. Altmann. Chicago: University of Chicago Press.

Washburn, S., and I. DeVore
1961 The social life of baboons. *Scientific American* 204: 62-71.

Washburn, S. L., and D. A. Hamburg
1968 Aggressive behavior in Old World monkeys and apes. In *Primates*, ed. P. Jay. New York: Holt, Rinehart and Winston.

Watson, A., ed.
1970 *Animal Populations in Relation to Their Food Resources: A Symposium of the British Ecological Society*. Oxford: Blackwell Scientific Publications.

Weber, I.
1973 Tactile communication among free-ranging langurs. *American Journal of Physical Anthropology* 38: 481-486.

Weir, B., and I. W. Rowlands
1973 Reproductive strategies of mammals. *Annual Review of Ecology and Systematics* 4: 139-163.

Williams, G. C.
1966a Natural selection, the cost of reproduction and a refinement of Lack's principle. *American Naturalist* 100: 687-692.

1966b *Adaptation and Natural Selection*. Princeton: Princeton University Press.

Wilson, E. O.
1971 Competitive and aggressive behavior. In *Man and Beast: Compar-*

ative Social Behavior, ed. J. F. Eisenberg and W. S. Dillon. Washington, D. C.: Smithsonian Institution Press.

1975 *Sociobiology: The New Synthesis.* Cambridge: Belknap Press of Harvard University Press.

Wilson, W. L., and C. C. Wilson

In press Species-specific vocalizations and the determination of phylogenetic affinities of the *Presbytis aygula-melalophus* group in Sumatra. *Proceedings of the Fifth Congress of the International Primatological Society.* Basel: Karger.

Wolf, K., and J. Fleagle

In press Adult male replacement in a group of silvered leaf-monkeys (*Presbytis cristata*) at Kuala Selangor, Malaysia. *Primates.*

Wooldridge, F. L.

1969 Behavior of the Abyssynian colobus monkey, *Colobus guereza*, in captivity. M. A. thesis, University of Florida.

1971 *Colobus guereza:* birth and infant development in captivity. *Animal Behaviour* 19: 481-485.

Woolfenden, G. E.

1975 Florida scrub jay helpers at the nest. *The Auk* 92: 1-15.

Yerkes, R. M.

1939 Social dominance and sexual status in the chimpanzee. Quarterly Review of Biology. 14: 115-136.

Yoshiba, K.

1967 An ecological study of Hanuman langurs, *Presbytis entellus. Primates* 8: 127-154.

1968 Local and intertroop variability in ecology and social behavior of common Indian langurs. In *Primates*, ed. P. Jay. New York: Holt, Rinehart, and Winston.

Zuckerman, S.

1932 *The Social Life of Monkeys and Apes.* London: Routledge and Kegan Paul.

Index